D1637267

The Sorcerer's Apprentice:

The French Scientist's Image of German Science
1840-1919

by Harry W. Paul, 1972

University of Florida Press / Gainesville / 1972

COPYRIGHT © 1972 BY THE STATE OF FLORIDA
DEPARTMENT OF GENERAL SERVICES

*Library of Congress
Catalog Card Number 77–178986
International Standard Book Number
0–8130–0347–4*

SERIES DESIGNED BY STANLEY D. HARRIS

PRINTED BY THE
STORTER PRINTING COMPANY
GAINESVILLE, FLORIDA

Acknowledgments

I wish to thank the University of Florida for its financial help during the preparation of the monograph. I am also grateful to Madame Antoine and Monsieur Daniel Langlois-Berthelot of the Archives Nationales and Madame Marchand of the Archives de l'Académie de Paris for making my archival work much easier than it would otherwise have been. My final thanks go to Diane Curry, who typed several versions of the manuscript, and to Jane Klinedinst and employees of Sponsored Research at the University of Florida, who typed the final draft.

TO MARI

Preface

In the late eighteenth and early nineteenth centuries France was the scientific leader and pedagogue of the world. By mid-nineteenth century the center of science was perceptibly shifting from Paris to a number of German centers, such as Heidelberg, Giessen, Göttingen, and Berlin, and, in some areas of science, to London and Edinburgh. From the 1830s on there were many warnings by French scientists, especially chemists like Dumas and Wurtz, that France was falling behind in the competition for scientific supremacy. But until the 1860s little action was taken to correct French material and institutional deficiencies, regarded as the cause of France's loss of supremacy. In the 1860s the French began to survey scientific developments in other countries, especially laboratory facilities in Germany, and to emulate them at home. This led to a tremendous expansion of the French scientific establishment and to a new period in the fertile history of French science, the period with which we associate Duhem, Poincaré, Lippmann, Langevin, Perrin, the Curies, and Le Châtelier. This brilliant Third Republican development, inspired, to some extent at least, by the ideas of competing for national survival against Germany and of maintaining France's position in the international scientific community, took place after the Prussian Army's defeat of the Second Empire in 1870–71. Believing the common but dubious argument that France had been defeated by German science, the Third Republicans, many of whom were scientists, were determined to prevent a repetition of such a catastrophe.

This essay deals with the French scientist's image of foreign, chiefly German, science. In it are traced the attitudes of French scientists, throughout the nineteenth century and past World War I, toward the development of science outside France. It was of obvious importance that France keep its pioneering role in science,

v

and closely related was the issue of the prestige of French science. Competition for foreign students, implying a *mission civilisatrice* for the French Isis, became serious in the late nineteenth century. France drew heavily on Russia, Rumania, and the Near East for foreign students in science and medicine. But because France was seriously concerned that large numbers of Americans went to Germany for their science and medicine, French scientists came to the United States and seriously attempted to make research and study conditions in France more attractive to Americans. Related to all these factors was the debate over the application of science to industry. With the coming of war in 1914 the scientist became the coryphaeus in the savage propaganda campaign unleashed by the intellectuals. All of the old issues were debated with renewed fury as France faced a threat to its national survival. Especially prominent in the debate early in the war was the physicist and philosopher and historian of science Pierre Duhem. After the war French scientists were powerful enough in international scientific organizations to ostracize the Germans. Although the English and the Americans shifted away from this boycott of the Germans, the German scientists themselves were so annoyed that they sabotaged the attempt of the Weimar politicians to promote international intellectual cooperation. French and German scientists thus remained fiercely nationalistic after the war and very sensitive to the idea of scientific pre-eminence and influence in the international community.

Contents

A la tête de la civilisation se placent l'Allemagne et l'Angleterre, qui par la découverte et le développement de l'Evolution, viennent de poser les bases d'un nouvel édifice de haute culture intellectuelle. —Haeckel, quoted in Denys Cochin, *L'évolution et la vie* (3d ed., Paris, 1888), p. 2.

La hegemonía de la ciencia alemana es una pretensión vanidosa.— José Aubin Rieu-Vernet, *¿La inferioridad de la ciencia francesa?* (Madrid, 1918), pp. 198–202.

Experience proves that it is rather the so-called "Intelligentzia" that is most apt to yield to these disasterous collective suggestions [psychoses of hate and destructiveness], since the intellectual has no direct contact with life in the raw, but encounters it in its easiest, synthetic form—upon the printed page.—Einstein to Freud, in Albert Einstein and Sigmund Freud, *Why War?* (Paris, 1933), p. 19.

1. Foreign-Related Issues
in French Science before 1914

ALTHOUGH WORLD WAR I provoked an outburst of pronounce-
ments by French scientists on German science unequaled since
1870, most French scientists had been concerned throughout the
nineteenth centry with scientific developments in Europe, especially
in Germany. It is true that a distorted image of German science is
found in French wartime propaganda, but it is impossible to judge,
without examining French scientific opinion on the same issues before World War I, whether these declarations were a significant
intellectual mutation or the logical culmination of a century of
reflection by successive scientific communities. This chapter is concerned with opinion on the chief issues related to foreign science
that occupied the attention of those French scientists who thought
about the problems resulting from the growth of science in neighboring countries. The prestige of French science was a constant
preoccupation of the French scientific mind in this period. Related
to this were the questions of whether French scientific development
kept pace with that in other countries and whether France's loss of
scientific pre-eminence was related to competition for foreign students. And as the nineteenth century ended, an old question was
debated with increasing passion: should priority be given to pure
rather than applied science and technology? Although all these issues were fiercely debated in World War I, as part of the matter of
national survival, their origins lie deep in the nineteenth century.

SCIENCE AND FOREIGN PRESTIGE

The importance of French science and maintenance of its reputa-
tion in Europe was a common theme throughout the nineteenth
century. The factual basis of this egoistic view can be found in the
earlier part of the century, stated in one of its extreme forms by the
great German chemist Justus von Liebig at the International Ban-
quet of Chemists in Paris in 1867: "I am going to propose a toast

1

to the memory of two of the greatest French chemists—of two of the founders of modern science, whose admirable works have never been surpassed, and still remain our models—of two *savants* who, as men, represent the most elevated qualities of the French nation. You will guess that I allude to Gay-Lussac and Thénard."[1] Writing to the vice-rector of the Sorbonne in the same year, the Dean of the Faculty of Sciences, the zoologist Henry Milne-Edwards, harped on the same theme: "La réputation européenne des savants qui occupent nos chaires fait la force de nos écoles scientifiques et peut être comptée comme une gloire de la France."[2] After the Franco-Prussian War, when, according to the chemist Henri Sainte-Claire Deville, France had been defeated by science, French scientists became more obsessed with foreign prestige. In 1872 the Spanish government named the mathematician Joseph Liouville "Commandeur extraordinaire de l'ordre de Charles III." Jules Simon, then Minister of Education, expressed his delight, and Liouville conveyed to Simon his pride at Spain's recognition of France's mathematical pre-eminence: "J'oserai ajouter que mon patriotisme est également sensible aux éloges que M. José Etchegaray (Ministro de Fomento) accorde au *Journal de mathématiques* dont je dirige la publication depuis près de 40 ans. Je suis heureux de voir qu'en Espagne du moins on reconnaît que dans les études mathématiques la France n'a pas cessé d'être sur le premier rang."[3]

At this point a brief note on the structure of the French university is necessary. The decree of December 28, 1885, established a sort of coordinating body for each university, a *conseil général des facultés*, made up of the rector, the deans of the faculties, the head of the school of pharmacy, and two professors of each faculty or school elected for three years by the faculty or school. The General Council published the list of courses and lectures, established the coordination necessary among studies, made sure that the courses contained the material necessary for examinations, gave its advice on the transformation or establishment of chairs, and advised the Ministry of Education on the division of the budget among facul-

1. *The Laboratory* 1 (July 20, 1867): 285. Liebig had worked with both in Paris.

2. *Registre des procès-verbaux, des actes et des délibérations de la Faculté des Sciences de l'Académie de Paris—Pièces annexes*, no. 2, pp. 10–11. Hereafter cited as *PV Sciences*.

3. *Archives nationales*, 87 AP 5 (*Fonds Jules Simon*), Liouville to Simon, December 12, 1872. Hereafter cited as AN.

ties. In addition to other administrative duties, it sent a general report to the ministry each year on the state and needs of the faculties. Another body, the Assembly of the Faculty, was made up of the teaching personnel, including *maîtres de conférences* and *chargés de cours*. Its chief job was to discuss the teaching program submitted to it each year by faculty members. The assembly also presented its candidates for dean every three years and elected two members to the *conseil général*. A third body, the *conseil de la faculté*, embodied the "personne morale" of the faculty as a public establishment. This important group was made up of the permanent professors of the faculty. It discussed the suppression, transformation, or creation of chairs and prepared the list of candidates for them; it managed the property of the faculty and handled its finances; and it presented to the ministry its own list of candidates for dean.[4]

In 1895 the mathematician Gaston Darboux, Dean of the Faculty of Sciences between 1889 and 1903, inaugurated for his faculty a custom of the Faculty of Arts, that of having a meeting when the term opened. In his address at that time, Darboux made explicit mention of the foreign students drawn by the prestige of France, especially in mathematics at the Sorbonne, and by a traditional attachment to France in the case of Rumanians and some Swiss. In the first year in the new Sorbonne quarters, for which the dean publicly thanked the government, the task would remain the same: "L'essentiel . . . est maintenant de mettre à profit tout ce qui nous est accordé et de travailler ainsi à la fois, et pour la science, et pour notre chère patrie."[5] Devotion to science and to France were complementary, and the idea of a possible conflict was quite remote. The philosopher Louis Liard, Directeur de l'enseignement supérieur, declared in 1896, after the paper shuffle reorganizing the University into universities, that although the universities should maintain intact the heritage of classical culture, they should also be open to all new sciences that could be for France a force and a defense in the struggle among peoples: "On a confiance qu'elles seront ce qu'on a voulu qu'elles fussent: des foyers d'indépendance, des ateliers de science, des écoles de patriotisme."[6] When the mathema-

---

4. See *La grande encyclopédie* 16: 1066–67, and Antoine Prost, *Histoire de l'enseignement en France, 1800–1967* (Paris, 1968), pp. 237–38.

5. *PV Sciences*, November 16, 1895, p. 206.

6. *Procès-verbal, Conseil supérieur de l'instruction publique*, July 1897.

tician and later politician Paul Painlevé gave a series of lectures on differential equations in Sweden in 1895, Darboux hailed the event as a triumph for French science: "M. Painlevé a porté au loin le renom de notre Université et de notre pays. . . . La Faculté des sciences est grandement honorée d'avoir été ainsi représenté."[7] The symbiosis of science and patriotism was more prominent in the scientific and educational establishment after the Franco-Prussian War than at any time since the Napoleonic period. It was especially characteristic of the Third Republic from 1877 to 1896, when the educational structure was overhauled and science faculties underwent considerable expansion. The survival of France, i.e., the success of the Republic, was made contingent on the development of science. It was necessary to participate in the international scientific movement, thereby maintaining or perhaps recapturing great power status, or risk degenerating into a stagnant fen. This was clearly put to the government early in the Third Republic by a commission headed by the chemist and powerful politician of French science Jean-Baptiste Dumas. This group had the task of judging various plans for the enlargement of the Sorbonne: "L'organisation des laboratoires de chimie, de physique et d'histoire naturelle, analogues à ceux que les universités de l'Allemagne ou de l'Angleterre possèdent, devient donc indispensable. . . . Comment maintenir la situation élevée de la France dans le mouvement scientifique qui agite le monde, si les nations rivales offrent à la jeunesse des secours dont parmi nous elle serait déshéritée."[8] It was only the most sublime example of a strongly established pattern and outlook when Pasteur rejoiced in his triumph over rabies as a triumph for France, all the more glorious since the victim came from Alsace: "Je suis bien heureux que ce nouveau succès soit dû à la France, et que le premier sujet humain chez lequel la rage aura été empêchée après morsure soit venu d'Alsace."[9]

Science as an instrument of foreign influence and prestige was

---

7. *PV Sciences*, December 18, 1895.

8. *Rapport adressé à M. le Ministre de l'instruction publique par la commission mixte chargée d'examiner les projets relatifs à l'agrandissement de la Sorbonne* (Paris, n.d.).

9. Pasteur à M. Liard, August 22, 1885, *Correspondence de Pasteur, 1840–1895*, 4 vols. (Paris, 1940–51), 4: 35. Pasteur was, of course, somewhat extreme in his reactions to Germany: see ibid., 2: 504–7, for an account of his return of his honorary degree to Bonn in 1871, and 4: 358–59 for his refusal to accede to Hermite's request in 1895 that he be put on the list of scientists recommended for the Prussian Order of Merit.

strikingly emphasized in the Dumas report to the Minister of Education in 1840. The report defended the practice of free admission of auditors to courses because of the significance for foreigners: "Car le concours de jeunes savants étrangers qui pressent aux portes de la Sorbonne, réagissant sur l'enseignement le maintient à une élévation qui fait la gloire de la Faculté."[10] Foreign universities tried to imitate the teaching of the Sorbonne, the report continued, without seeing that it was dependent on the very liberality of that education—that it was the natural product of an enthusiastic and enlightened audience reacting to professors worthy of its confidence. Naturally, the faculty knew that the force and life of its teaching and its prestige in the eyes of Europe came from the seriousness and profundity of its teaching. French scientists maintained the image of themselves as the scientific pedagogues of Europe. Up to that time the material deficiencies becoming evident in French science had had, in the opinion of the French at least, little effect on the scientific balance of power. But the French scientists wanted the material deficiencies corrected to make sure that their pre-eminence would not be threatened by foreign opulence.

Comparisons with Foreign Science

By the 1830s the French were conscious of scientific developments in neighboring countries that were, in many areas, a possible threat to French leadership. In 1837 a government commission, in setting forth a group of recommendations for the whole university, showed that French scientists were keenly aware that France could not rest on its scientific laurels. The commission recognized that the immense development of the mechanical arts in England had given her a clear superiority over France in this area.[11] In 1840 the report of Dumas to the Minister of Education emphasized the opulence of foreign universities as compared with the Sorbonne: "Dans un moment où l'Angleterre, la Prusse, les états de l'Allemagne, la Russie, font les plus grands efforts pour donner à leur enseignement universitaire un luxe inaccoutumé, vous ne voudrez pas, M. le Ministre, que la Faculté des sciences demeure en arrière de ce noble mouvement de l'esprit humain." Embarrassing questions were asked. Why should the Faculty of Sciences in Paris be inferior materially to the universities of London, Edinburgh, and Berlin? Why were

10. *PV Sciences*, July 1, 1840, pp. 111–16.
11. Ibid., December 6, 1837, p. 81.

there no useful models in mechanics like those London gave to its professors? Why was natural history deprived of needed materials?[12] Although only separated by three years, the 1840 report contained an urgent note not present in the nearly complacent report of 1837. Material deficiencies in French astronomy were lamented by the astronomer Urbain Leverrier in 1849, when he called for the purchase of good instruments and other items necessary for physical astronomy.[13] By 1857 he was complaining of the absence of astronomical observers in France as compared with the large number of German scientists in this specialty. Leverrier thought that to follow a suggestion of the mathematician Michel Chasles, that mathematical astronomy and physical astronomy be taught separately, would only make matters worse.[14]

But it is hard to find much interest generated by the reports and complaints of scientists in the first half of the nineteenth century. Only in the Second Empire was there a serious effort to implement some of the costly recommendations of the scientific community. Even in the early 1850s the government seemed unmoved. In a speech in 1852, the Minister of Education, Hippolyte Fortoul, gave his opinion that the supremacy of France was assured by the suitability of French as the language of science. This view existed in subdued form throughout the century, but it was only after 1914 that it developed into a vitriolic attack against the obscurity of German:

> Notre langue ne semble-t-elle pas aussi particulièrement conviée à la culture des sciences? Sa clarté, sa sincérité, son tour vif à la fois et logique, qui substitue partout avec rapidité l'ordre de la pensée à l'ordre de la sensation, ne l'ont-ils pas destinée à être non-seulement leur instrument le plus naturel, mais même leur guide le plus utile? Ses beautés, toutes de vérité et de raison, ne sont-elles pas la parure la plus heureuse qu'elles puissent revêtir? Si Descartes, Pascal, Fontenelle, Buffon ont puisé dans les sciences la grandeur régulière, la profondeur solide, la délicatesse, l'éclat qu'elles ont tour à tour prêtés à la langue française, n'est-ce point pour qu'elle rende aux sciences les services qu'elle en a reçus? N'aurions-nous enfin une langue habile à dessiner avec une pureté exquise les

12. Ibid., July 1, 1840, p. 116.
13. Ibid., January 23, 1849, p. 111.
14. Ibid., June 23, 1857, pp. 232–33.

contours des choses, que pour lui interdire les sujets où elle peut déployer avec le plus d'utilité sa précision admirable? N'aurions-nous un idiome excellent, entre tous, à montrer la force de l'entendement toujours présente dans les images mêmes des objets les plus sensibles, que pour lui refuser de nous donner le témoignage le plus décisif de l'empire de la pensée sur la matière? Dans le siècle où l'homme a su réduire l'air, le son, la lumière à ses mesures, et soumettre l'invisible et l'impalpable à ses observations, devons-nous craindre qu'il oublie sa dignité et qu'il abaisse sa prééminence en cultivant les sciences qui lui ont permis de fournir les exemples les plus fameux de la supériorité de son esprit?[15]

By the 1860s there had emerged a somewhat different approach to the study of foreign scientific facilities. This is perhaps best exemplified by the famous reports on German laboratories by Adolphe Wurtz, Dean of the Faculty of Medicine in Paris between 1866 and 1875, at which time he became Professor of Organic Chemistry at the Sorbonne. He had been asked by Victor Duruy, Minister of Education, for information on "practical chemistry" and on the state of laboratories in foreign countries. Wurtz signaled the tremendous progress that Germany had made in chemistry since Liebig had opened his laboratory a generation before in Giessen:

Depuis l'époque déjà éloignée où M. Liebig rassemblait à Giessen des élèves venus de tous les pays du monde, et fondait une école justement célèbre, les études de chimie ont pris un grand essor en Allemagne. De vastes laboratoires ont été construits à Giessen, à Heidelberg, à Breslau, à Göttingen, à Carls-ruhe, à Greifswald. De nombreux travaux, de belles découvertes ont été le fruit de ces utiles créations. La science en a profité; l'Allemagne s'enorgueillit à juste titre et redouble d'efforts pour mettre l'enseignement de la chimie pratique à la hauteur des progrès et des exigences modernes.[16]

After furnishing details of construction and cost, Wurtz concluded that comparison of German and French achievements was quite unfavorable to French laboratories. It was necessary for the minister to take some action to improve the situation:

15. *Discours* (August 12, 1852), AN, 246 AP 19 (*Fonds Fortoul*). Author of *De l'art en Allemagne*, 2 vols. (Paris, 1841–42), admirer of the classics, bête noire of the liberal academics, Fortoul did take an interest in science and tried to get good scientists appointed to the provincial faculties.
16. Wurtz to the Minister of Education, December 10, 1864, AN, F17 4020.

Il s'agit là d'un intérêt de premier ordre, de l'avenir de la chimie en France. Cette science est française et Dieu ne plaise que notre pays s'y laisse devancer. Et le danger existe, car on peut affirmer que le mouvement scientifique, tel qu'il se manifeste par le nombre des découvertes et des publications utiles s'est prononcé davantage, dans ces dernières années, en Allemagne qu'en France. L'impulsion est partie de notre pays; mais elle s'est propagée avec une grande puissance au delà de nos frontières.[17]

In 1868 Wurtz became lyrical about "ce beau monument élevé à la science," the laboratory at Bonn. Noting a certain luxury in the laboratory, he thought it justified: "Mais ce luxe est digne d'un grand peuple qui comprend le rôle de la science et l'importance des choses de l'esprit dans la marche de la civilisation."[18] His sadness at the absence of similar facilities in France increased when he learned of the plans to build a new Faculty of Medicine at Bonn costing over two and a quarter million francs. Wurtz' conclusion, evidently accepted by Duruy and the government of the Second Empire and embodied in the new science policy that created the Ecole pratique des hautes études in 1868, was that immediate action had to be taken if France were not to be left far behind in the competition for scientific supremacy. The eclipse of the Faculty of Medicine in Paris by that of Bonn would be only the most glaring example. Wurtz noted that public opinion was also beginning to be concerned with laboratories. The needs of science and the requirements of higher education had thus merged with the highest political interests. Wurtz' final warning: "Je me crois . . . autorisé à déclarer . . . que la situation qui est faite actuellement en France à l'Enseignement scientifique ne peut se prolonger sous peine de mettre en péril la suprématie de la science française et l'avenir intellectuel de notre pays."[19]

At the end of 1867 Duruy showed that the government was fully aware of the challenge to the scientific pre-eminence of France and of the need to eliminate the material deficiencies plaguing French science:

> M. le Ministre termine en insistant sur les lacunes que présentent encore l'enseignement supérieur. A cet égard, la prépondérance de la France n'est plus incontesté; si elle n'a

17. Ibid.
18. Wurtz to the Minister of Education, April 8, 1868, ibid.
19. Ibid.

reculé sous aucun rapport, ses rivaux ont fait des sacrifices et des efforts qui doivent la préoccuper. L'insuffisance des ressources budgétaires n'a pas permis à la France de développer et de renouveler . . . le matériel scientifique de nos facultés. Les traitements sont insuffisants; les laboratoires manquent, et c'est à grand peine et grâce à des expédients, qu'on a pu, cette année, créer à la Sorbonne, dans une arrière-cour le seul laboratoire de physique, vraiment digne de ce nom, qui existe en France. Il en faut d'autres pour toutes les branches des sciences et pour tous les hommes éminents au Collège de France, au Muséum, à la Faculté de Médecine, à l'Ecole normale, partout enfin où un homme supérieur aura besoin, pour faire avancer la science, de trouver autour de lui les instruments les plus perfectionnés et les auxiliaires les plus intelligents.[20]

Not all French scientists were as impressed as Wurtz with German scientific developments. Marcelin Berthelot had visited Heidelberg in 1858 and was by no means in favor of radically changing the direction of French science toward emulation of Germany:

Ici, écrivait-il à Renan, je vois beaucoup de choses qui me montrent plus clairement, par contraste, les avantages du système français. Je cause souvent avec les privat docent scientifiques: leur vie est bien plus misérable que la nôtre et les instruments de travail leur manquent bien plus complètement. Le peu d'indépendance qu'ils pourraient avoir en compensation est plutôt virtuel que réel, car l'homme très pauvre ne peut guère être regardé comme vraiment libre. Quant aux professeurs, je ne sais si la direction de 50 élèves assujettis à exécuter un cours régulier de 150 manipulations toujours identiques n'équivaut pas aux examens et aux doubles fonctions pour la neutralisation de l'activité scientifique. Un seul avantage existe ici: l'absence des préoccupations ambitieuses qui perdent tous nos savants dès qu'ils arrivent à l'âge mûr. Mais, en retour, l'horizon est étroit et l'on ne peut guère acquérir ce sentiment vif et général des choses que l'on trouve à Paris. Toutes choses, je crois, doivent être vues à la fois de près et de loin: ici, je suis bien plus optimiste relativement au système français dont nous connaissons si bien les énormes inconvénients. . . .[21]

Even in 1882 Berthelot wrote a letter to the linguist Michel Bréal comparing the material situations of French, English, and German

20. *Conseil impérial de l'instruction publique*, December 9, 1867, AN, F17 12958.
21. Léon Velluz, *Vie de Berthelot* (Paris, 1964), pp. 45–46.

professors, but he warned against the game of comparing the worst features of the French system with the best features of the English and German systems: "Je crains que nous ne voyons toujours l'âge d'or à côté de nous et l'âge de fer chez nous. Certes, il n'est pas inutile de faire valoir la situation des professeurs allemands pour améliorer celle des professeurs français vis à vis les Chambres et le public, mais il ne faut pas, non plus, être dupe de cette comparison, où l'on oppose les côtés defectueux de notre système aux côtés brillants de celui de nos voisins."[22]

The type of report done by Wurtz became quite common during the Third Republic. This activity indicates that French science kept abreast of developments elsewhere, especially in Germany, and moved rapidly to recommend innovations in France. During the latter part of the nineteenth century, many of the demands of the French scientific community were met by the government. Deficiencies noted were nearly always material ones. Such is the case of the physiologist Charles Richet's report in 1880 on the organization of physiology laboratories in the Netherlands (Place at Amsterdam, Donders and Englemann at Utrecht, and Heynsius at Leyden). Richet noted that electro-physiology was practiced much more in Holland and Germany than in France, where inferiority in this area stemmed mostly from poverty in the required instrumentation, itself a result of a lack of money. (Richet observed that Dutch students were generally rich enough to spend between 2,500 and 3,500 francs per year. One compulsory personal expense was a microscope, costing between 175 and 225 francs.) As a result of his study, Richet recommended two chief improvements for French laboratories. First, there should be a large room reserved for physiological chemistry, the precise procedures of which could be followed only without interruptions from other laboratory activities. Much equipment had to be furnished: "Sinon on restera bien loin en arrière des laboratoires allemands et hollandais, où les recherches chimiques exactes occupent tant de place en physiologie." Second, the post of mechanic-workman should be introduced. Such a person was indispensable for repairing and constructing instruments and apparatus. In Amsterdam he was regarded as vital to the functioning of the laboratory: "Cet ouvrier est notre bras droit, et il nous serait impossible de nous en passer." On

22. This quote is taken from an old catalog advertising the sale of several Berthelot letters. M. Daniel Langlois-Berthelot kindly called my attention to it.

the other hand, Richet found that the separation of general anatomy and physiology, as practiced in France, was preferable to mixing these subjects, as in the Netherlands. But much money and effort were necessary for Paris to equal any of the three laboratories visited.[23] Emphasis was not always on material deficiencies, and it shifted, especially after the end of the Second Empire, to the absence of certain scientific subjects in the curriculum.

Reference to the structure of science in other countries was also a useful weapon in getting change in the science faculties. In 1874 Wurtz got Pasteur's support for his proposal to offer a *cours complémentaire* in organic chemistry in the Faculty of Sciences at the Sorbonne. Wurtz' letter to the faculty argued that the absence of such a course created a grave lacuna in the teaching of the faculty.[24] In 1876 Joseph Boussinesq, of the Faculty of Sciences at Lille, used, as part of his maneuver to get to the Sorbonne, the absolutely correct argument that higher education as organized in Paris had a regrettable gap: the application of mathematical analysis to terrestrial mechanics and especially to hydrodynamics, to the theory of static or dynamic resistance of solids, etc. (". . . Il serait honteux pour la France de laisser la meilleure part aux Allemands ou aux Anglais"). In a second letter to the Minister of Education four months later, Boussinesq strengthened his emphasis on developments in Germany, where all the important universities carried the type of courses missing in Paris.[25] In the republic of professors the complaints of both Wurtz and Boussinesq were recognized as legitimate: Wurtz got his course in organic chemistry, and eventually Boussinesq provided higher education in Paris with what he had been doing in mathematics and mechanics for many years in Lille.[26]

As already indicated, comparisons of France were made not only with Germany, but frequently with other countries, chiefly England, Italy, and Russia. The mathematician Victor Puiseux pointed out

23. Charles Richet, "Rapport sur l'organisation des laboratoires de physiologie des Pays-Bas," *Archives des missions scientifiques et littéraires*, 3d ser. 6 (1880): 289–300.

24. *PV Sciences*, February 24, 1874, p. 351.

25. Letters of J. Boussinesq to the Minister of Education, 1876, AN, F17 13074.

26. Another example of using a comparison of the Sorbonne and a foreign university faculty was Armand Gautier's attempt in 1878 to have a chair of general toxicology created for him in the faculty. AN, F17 13074.

in 1877 that, although the Sorbonne had two chairs in astronomy, certain recent discoveries in stellar astronomy, in the constitution of the sun and planets, and in celestial photometry could not be treated in existing courses. Puiseux thought that there would be general interest in this new science in France, although there were few adherents compared to the number in England and Italy, where this field was cultivated by a large number of astronomers. At this suggestion, Chasles reminded the faculty that he had already asked for a chair of geometry, as already existed in many foreign universities. Four such chairs had been recently created in Italy.[27] But there was not the same sense of urgency in these comparisons as in those with Germany. Frequently, the developments in Germany seemed to have more serious economic and political implications.

Inevitably, some questions were raised about the effects of imitating other countries. An interesting debate arose in the council of the Sorbonne Faculty of Sciences in 1882 concerning the vacant chair of mathematical physics and calculation of probabilities. The French were clearly aware that radical new developments had taken place in certain areas of physics and chemistry. Some members of the council argued that the introduction of German science into the program of mathematical physics would reduce purely mathematical considerations. The mathematician Ossian Bonnet did not agree but admitted that the "new science" would probably require for the solution of its problems the creation of new mathematical procedures as happened when mathematical theories were created for heat and elasticity. To calm anxious minds, Bonnet assured the council that the new science was French in essence: "Ouvrez la philosophie naturelle de Thomson et les leçons de Riemann, celles de . . . Kirchhoff . . . de Christoffel, de Beltrami, vous ne trouverez pas autre chose que la science française." The sacred names of the famous French physicists and mathematicians Joseph Fourier, Ampère, Denis Poisson, and Gabriel Lamé were invoked to assure the doubters.[28] This was not an idle discussion: the appointee was

27. *PV Sciences*, February 21, 1877, p. 30.

28. Ibid., November 14, 1882, pp. 57–58. The engineer-chemist Georges Lemoine, in "L'évolution de la chimie physique," *Revue des questions scientifiques* 73 (1913): 62, emphasized the French "origins" of physical chemistry: ". . . Berthollet la première origine de la chimie physique. . . . Mais ce sont Berthelot et Henri Sainte-Claire Deville qui sont les fondateurs de la chimie physique: Deville avec son idée géniale de la dissociation, Berthelot avec son travail sur l'éthérification."

Gabriel Lippmann, who had taken a doctorate at Heidelberg and had worked in Kirchhoff's laboratory. The faculty took seriously the idea that science had certain national characteristics, implying that French science had certain values that had to be preserved from outside corruption.

The problems evolved more clearly in a discussion in 1898 of the nature of French science. The occasion was the need to choose the best scientist available to teach a course in physical chemistry. Louis Raffy, Anatole Leduc, Pierre Curie, J. Guinchant, Jean Perrin, A. Ponsot, and Georges Charphy were the leading candidates. Darboux, the dean, pointed out that the purely mathematical direction that Gustave Robin had given this course could no longer be maintained. Emile Duclaux, biochemist and head of the Pasteur Institute, added that physical chemistry was a special science demanding aptitudes different from those needed for physics. It required modes of reasoning to which the French were not accustomed and demanded a special type of penetrating mind. Rather than appoint a man who had already determined his form of reasoning, which would prevent him from easily adapting to the forms of the new investigation, Duclaux would take a young scientist of independent mind who had withdrawn from the French physics laboratories for a few months' stay in Germany. The physicists Edmond Bouty and Henri Pellat disagreed. They thought that it was precisely by the use of vague reasoning that German and Dutch scientists and especially the Swedish scientist Svante Arrhenius, the collaborator of J. H. van't Hoff and Wilhelm Ostwald, had filled this branch of science with obscurities and incoherence. The role of the French mind should be to clarify: "N'est-il pas le rôle d'un esprit français d'y apporter la lumière et de faire disparaître toutes ces hypothèses hasardées qui s'évanouissent quand on s'efforce de les étudier et de les préciser?" The chemist Charles Friedel thought Duclaux' idea to be dangerous; he feared that the man who returned from Germany would be more a German physical chemist than a French one. He did not fear to see the clarity of the French mind applied to this area. The candidate should have a certain authority and solid knowledge; a young man would be too susceptible to foreign influence. Lippmann, who had spent much time in Germany and was supporting Curie for the job, agreed that contemporary physico-chemical doctrines carried the mark of the countries in which they were born. Finally, Perrin got thirteen votes, Curie seven, and Ponsot three.[29] The theory of national styles of scientific thinking, made familiar by Duhem's well-known statement, was common among French scientists in this period and exercised a definite influence on decisions that vitally affected the course of science. With Perrin's appointment, of course, the Sorbonne had a man who was in the forefront of the same type of research as other European physical chemists.

By 1900 a quite different view of the importance of physical chemistry was common in France, partly due to the scientific work and writing of Pierre Duhem:

> Le développement prodigieux de la Chimie-Physique restera sans doute l'une des marques, et non la moins belle, du XIX⁰ siècle finissant. . . .
> Les principes sur lesquels repose la Chimie-Physique menacent d'une révolution les vieux systèmes cosmologiques; les conséquences de cette science font prévoir des bouleversements dans les procédés de l'industrie chimique. Une nation qui aspire à l'hégémonie à la fois dans le monde intellectuel et dans le monde économique devait, nécessairement, revendiquer la première place en Physico-Chimique et, pour parvenir à cette place, s'assurer le concours d'un guide capable de l'y conduire. L'Allemagne ravit donc M. J. H. van't Hoff à Amsterdam et l'amena à Berlin.[30]

While he himself "languished" in Bordeaux, Duhem did not see any evidence in France of a movement comparable to that in Germany. Quite the contrary: according to Duhem, some members of the scientific establishment in Paris, chiefly Berthelot, were stifling the type of research Duhem thought important. But Berthelot was only one of the large number of powerful scientists in Paris, and, to some extent, this judgment reflects Duhem's hostility towards Berthelot.

---

29. *PV Sciences*, April 11, 1898, pp. 256–57.
30. P. Duhem, "L'œuvre de M. J. H. van't Hoff. A propos d'un livre récent" (*Leçons de Chimie-Physique*, 2 vols. [Paris, 1898–99], translated by M. Corvisy, Professeur agrégé au Lycée de Saint-Omer), *Revue des questions scientifiques* 47 (1900): 6. See also Duhem, *Une science nouvelle, la Chimie-Physique* (Paris, 1899). Duhem added that the Dutchman had been received into the Berlin Academy of Sciences, was an Ordinary Professor at the University of Berlin, had his own laboratory, and was head of the Institute of Physics of Charlottenbourg. The first number of the *Zeitschrift für physikalische Chemie*, which was edited by Ostwald, appeared in February 1887 with the name of van't Hoff as co-founder and editor.

COMPETITION FOR FOREIGN STUDENTS

In the international competition for students, France had little success in attracting Americans to study under the world-renowned French scientists of the second half of the nineteenth century. It is significant that the French scientific community recognized this, worried about it, and tried to make certain reforms that would attract more foreigners, especially Americans, some of whom perhaps were thinking of working in France. In 1895 the opinion in the Sorbonne Faculty of Sciences was that more American students would study in France if they could obtain a degree, as was the custom in some foreign universities.[31] This information was used in reforming the license program. In the same year the French introduced the *doctorat d'université*: this required the same high standards and originality in research as the *doctorat d'état* but made it

31. According to *The Nation* (New York), in 1896 there were 2,192 foreign students in Germany. Of these, 515 were Russian, 492 American, 316 Austro-Hungarian, 283 Swiss, 139 English, 96 Belgian, 56 French, 44 Dutch, 34 Italian, 31 Swedish and Norwegian, and 25 Rumanian. In 1896 there were over 14,000 students registered in the Sorbonne (4,518 in law, 5,445 in medicine, 1,684 in letters, 500 in science, 1,802 in pharmacy, and 55 in Protestant theology). Of the total, 1,400 were not of French nationality. There was a heavy foreign registration in medicine, much to the alarm of French medical students who organized to prevent foreign doctors from practicing in France. Most of the foreign medical students came from Russia, Turkey, Egypt, Bulgaria, Rumania, and Greece. *Revue de l'institut catholique de Paris*, no. 1 (1896), p. 362; no. 2 (1896), p. 73. The total number of foreign students at the Sorbonne in 1889–90 was 986. In two years the number had risen to 1,251. *Enquêtes et documents relatifs à l'enseignement supérieur*, 41 (1891): 11, and *PV Sciences*, December 18, 1895, p. 35. See also Maurice Caullery, "La recherche scientifique aux États-unis," *Revue internationale de l'enseignement* 71 (1917): 161–80, 241–67.

NUMBER OF FOREIGN STUDENTS AT THE SORBONNE, 1889–90

	LAW	MEDICINE	SCIENCES	LETTERS	PHARMACY	TOTAL
Germany	7	6	1	9	–	23
North America	8	144	7	8	–	167
England	5	45	4	9	–	63
Austria	5	6	–	5	1	17
Spain	1	28	–	–	3	32
Italy	2	11	1	–	1	15
Portugal	–	17	–	–	–	17
Rumania	47	77	13	11	1	149
Russia	9	247	18	15	3	292
Serbia	8	18	–	2	2	30
Switzerland	5	24	1	6	1	37
Turkey	19	69	1	–	4	93

possible for foreigners to be credited for their work outside France, thus avoiding the difficult and time-consuming chore of getting the licenses required for the *doctorat d'état*. Unfortunately, foreigners quickly concluded that the new degree was inferior to the *doctorat d'état*, and the reform never achieved its goal.

A long discussion about American students took place in 1902, after Jacques Hadamard, a brilliant mathematician, made a trip to the United States and reported to the Sorbonne Conseil de l'université. Hadamard talked with American professors about what prevented their sending students to Paris. The first reason given was that the French published the course programs too late in the year for Americans to plan their studies. To eliminate this difficulty, the programs were subsequently published in April and sent to American universities. A second complaint concerned restricted access to and insufficient working hours of libraries. The same problem existed for the laboratory libraries.[32] During the discussion, Alfred Giard, a biologist, indicated that the situations in Germany and Belgium were better, and Lippmann stated that he thought there should be a separate science library. The third problem indicated in Hadamard's report was that French professors did not aid the students as actively or as effectively as the German professors.

The mathematician Emile Picard, who favored more help for doctoral students, gave a long analysis of the differences between the French and German handling of these students. French doctoral candidates had to produce a more substantial thesis than German students, despite the fact that French professors helped the students less. The German professor was the real author of the thesis—he gave the subject and guided the work all the way to completion. Picard argued that the French had never proceeded in this manner. The candidate had to show initiative and make his work a personal effort. In addition, in Picard's opinion, the German professors were more familiar with students and welcomed them as part of their families.[33] This intimacy did not exist in France, especially in math-

32. This was probably a European problem: in the 1870s Americans working in Friedrich Wöhler's chemistry laboratory in Göttingen got Wöhler's permission to work in the laboratory during Christmas vacation, although he refused to permit Germans to do the same. H. S. van Klooster, "Friedrich Wöhler and His American Pupils," in Aaron J. Ihde and William F. Kieffer, eds., *Selected Readings in the History of Chemistry* (Easton, Pa., 1965), p. 29.

33. This was not true of Liebig; but Wöhler was quite fond of his American students. Ibid., and Klooster, "Liebig and His American Pupils," ibid., pp. 18–22.

ematics where, because there was no laboratory work, there was less contact than in the other sciences. Henri Moissan, professor of chemistry and head of the practical chemistry laboratory, who had been to the United States, agreed with Picard's analysis and pointed out that German professors actively sought students in the United States.

Albin Haller, who had left his chemistry institute in Nancy in 1899 to come to the Sorbonne, pointed out that most American science professors had studied in Germany; it was natural that they should send their students there. The German laboratories were organized differently, with the professor always present to guide the student's work.[34] Haller thought that the work of the German students was always more specialized, making it possible, for example, for the creator of a beautiful organic synthesis, a "doctor in chemistry," to use potash but not to know how to prepare it. The Germans were less demanding of their students, thus allowing them to finish their doctorates faster. But Haller emphasized that the German laboratories were better funded and had more personnel. There were from 20 to 50 per cent more doctors available than in France. Giard noted that the students of all important nationalities could find persons speaking their languages in the German laboratories, whereas in France the situation was hopeless. (This was certainly a change from the days of Liebig in Giessen, when arriving students were given only two or three days to converse in English.) Lippmann felt that it would not be easy to apply the German system in France because of the independence of the French student, less susceptible to discipline than foreign students. Most French students wanted to pick their own topics. Moissan pointed out an important moral factor: the University of Chicago had wanted to send some of its students to Paris but had decided against it after getting alarming information on material and moral conditions in the modern Babylon. The situation in Paris was very different from that of a typical American university; there were no *maisons de famille* where the student could get moral and religious training as well as an education—altogether, a rather depressing picture.

34. Wöhler left this mostly to his assistants, at least in his later years, when his laboratory had a large number of students: ". . . I enjoy lecturing and abhor laboratory periods. One might like the contact with a few intelligent people but not the company of 75, most of them incompetent." Klooster, ibid., p. 23.

The biologist Yves Delage summed up the differences between France and Germany from the viewpoint of the foreign student. First, there were French deficiencies in faculty equipment and personnel, although these could be remedied by the university. Second, there were undesirable living conditions, though these could be improved. But meekness would never become a characteristic of the Parisian student: "Paris transformera toujours les étudiants à son image avec l'indépendance de pensée et de conduite qui caractérise le milieu." Third, German professors helped students with their work to the point of lightening the burden of the student's research, a procedure which the French should not imitate.

In the next faculty meeting the dean asked for research on methods of promoting contact between students and professors. Most professors did meet the students outside the large public lectures, especially in the laboratories. Since the faculty wanted to do as much as possible to attract foreign students, its discussions returned to some of the points already treated, chiefly Haller's complaint of a lack of resources in Paris. In writing his report on the Chicago exposition, he had had to borrow books from Nancy, and in his research in chemistry he was often forced to get work done in Nancy, where he had had a credit of 75,000 francs, five times as much as in Paris. But Picard rightly pointed out that there was the same lack of foreign students in the Faculty of Letters, where laboratories did not play any role. After some talk about the lack of university sports facilities, the discussion moved to the fact that many Americans were not prepared to follow Sorbonne courses. Gabriel Koenigs, in mechanics and mathematics, believed that they found French teaching too abstract and theoretical. Haller did not think it could be made more practical because France lacked the developed scientific and industrial movement of Germany. Bouty was convinced that better organized laboratories would attract more foreign students, but Hadamard was skeptical. In mathematics, the situation was reversed, since nowhere could one find the same mathematical galaxy as at the Sorbonne. Yet the Americans stayed away. Picard also remarked that Americans did not come to France to learn to be manufacturers and businessmen, although some came to finish their education on a certain intellectual note, as, e.g., those who came to the Ecole des beaux arts. The dean drew attention to the unstructured nature of American education, where there was little concern with course coordination or with the preparatory knowledge of university students. Thus a professor would deal with the theory of forms before an audience which did not know differential calculus. There was a great variety of universities with differing standards. Secondary and higher education were frequently mixed without regulations, and programs were bewildering. The French had enlarged the framework of their teaching and increased the professors' independence in their choice of subjects, but it was impossible to go further without deforming the structure of French education. As Picard kept emphasizing, the only thing to do was to increase professor-student contacts. A Comité de patronage des étudiants étrangers already existed.

The discussion ended with a consideration of the doctorate. Hadamard indicated that part of the problem was that American professors generally regarded the *doctorat d'université* as a cheap degree, although some had come to see its value. Some even regarded the German degree as much weaker. Haller noted that in Germany the famous chemist Emil Fischer had lamented the weakness of the German degree and had called for requiring the candidates to have some preliminary achievement analogous to the French license. But Haller thought the two doctorates an inconvenience. No one wanted to change the French *doctorat d'état*, especially when some Germans saw its virtues and thought it wise to copy some of its features. The faculty decided to maintain the *doctorat d'université* at the same level as the *doctorat d'état*. There would be leniency in dispensing with the two certificates required for the *doctorat d'université*, the value of which would be made clear by publicizing that the only difference between the degrees was the easier access to the *doctorat d'université*. This information would be sent to the American universities with descriptions of each of the Sorbonne laboratories.[35]

French science faculties were keenly aware of competition between themselves and the Germans over the training of foreign students. When these scientists returned to their native universities, they were, the French hoped, agents of a French scientific *mission civilisatrice*. The French position was strong in Rumania, where the *doctorat d'état* was highly prized and legally required in certain appointments. The importance of the Rumanian contingent of students in France had always been recognized by the educational establishment. When the Rumanian Eugène Néculcéa defended his

35. *PV Sciences*, January 21 and 24, 1902, pp. 348–60.

thesis (*Recherches théoriques et expérimentales sur la constitution des spectres ultra-violets d'étincelles oscillantes, théorie interférentielle des appareils spectraux*) in 1906, the inspector present made clear in his report the significance of the event: "Ce travail magistral fait honneur à la science française qui l'a inspiré, et il consolidera notre influence scientifique en Roumanie, dont les étudiants sont toujours nombreux et assidus dans [nos] établissements d'enseignement supérieur."[36]

A competitive situation also existed in Switzerland, where the French were very sensitive to the need to maintain their position. In 1896 Jean-René Thomas-Mamert, professor in the cantonal University of Fribourg, defended his thesis (*Sur quelques aminoacides non-saturés*) and received the mention "très honorable," although he had not obtained decisive results on certain points. The jury and the inspector were confident that he would honor French science and maintain in this friendly neighboring country the French influence then so contested there.[37] In 1919 there was discussion in the Sorbonne of representation of the Faculty of Sciences at a conference of Swiss universities. It was recognized that difficulties would arise if enemy nations were represented. But a letter of the rector of the University of Lausanne to the Directeur de l'enseignement supérieur declared that there would be no German delegates. Bernard Bouvier, a professor at Geneva and a graduate of the Normale, had also visited the dean at the Sorbonne. From all this it was clear that if the Sorbonne faculty did not send representatives, the Germans would take advantage of their absence to argue that the French had no interest in Switzerland and would try to better their position there. A reading of the list of delegates revealed that they were all authentic Swiss. The only question was whether the Ministry of Education or of Propaganda would pay the delegates' expenses.[38]

THE DEBATE OVER PURE VERSUS APPLIED SCIENCE

A considerable number of French scientists have always been devoted to the practical or applied aspects of science. In the 1837 commission report there was clear recognition of the interaction of

36. AN, F17 13248.
37. Ibid.
38. *PV Sciences*, May 2, 1919, pp. 171–72. See Abel Rey, "Les relations universitaires franco-suisses," *Revue internationale de l'enseignement* 72 (1918): 19–31.

the two areas and the dependence of the practical segment on the more theoretical teaching of the faculty:

> ... En ce qui concerne les arts chimiques ... nos manufactures jouissent d'une supériorité incontestée ... ce qui s'explique ... par l'éclat que l'enseignement de la chimie en France conserve depuis 40 ans, et en particulier par le succès soutenu des cours de chimie de la faculté.
>
> La réaction de l'enseignement des sciences sur la pratique des arts industriels est donc à la fois prompte et efficace en France. Abandonner le perfectionnement de nos arts mécaniques à la pratique seule, c'est donc agir contrairement aux précédents et peut-être même contrairement au tour d'esprit national.[39]

Throughout the nineteenth century it is easy to detect tension between the more theoretically and more practically oriented of the faculty. The division sharpened, especially in the first decades of the twentieth century, but it had existed early in the nineteenth century. In 1834 the physicist-chemist Pierre-Louis Dulong argued that, given the state of French industry, it was of the greatest importance that the Faculty of Sciences have a chair in practical mechanics. C.-S.-M. Pouillet, who had too much material to cover in physics, supported Dulong's opinion that the Ministry of Education should be asked to create such a chair. But the mathematician Silvestre-François Lacroix thought that a course in elementary mechanics was not advanced enough to be taught in the Faculty of Sciences. Poisson believed such a course useful but that it should be taught in the Conservatoire des arts et métiers.[40]

French interest in technical training developments in German schools goes back at least to the 1830s.[41] After the Franco-Prussian War, awareness of this progress penetrated the higher councils of education.[42] By the 1890s the provincial science faculties were in-

39. PV Sciences, December 6, 1837 (Commission pour examiner quelles seraient les améliorations qui pourraient être apportées dans l'enseignement des Facultés des sciences: Libri, Mirbel, Dulong, Beudant, Dumas).

40. Ibid., February 11, 1834, pp. 40–41. Poisson also argued that Biot's proposed course in mechanical physics should be given in the Collège de France, which existed to deal with "les parties élevées de la science."

41. The deputy, journalist, and suppléant of Guizot, Saint-Marc-Girardin, drew French attention to some of the deficiencies in technical and vocational education in his study De l'instruction intermédiaire et de son état dans le Midi de l'Allemagne, 2 parts (1835–39). See also Victor Cousin's reports.

42. See, e.g., Programmes de l'école normale de l'enseignement secondaire

troducing courses related to local industry and agriculture.[43] In 1895 the implications of this growing rapport between science and industry were discussed in the Sorbonne Faculty of Sciences. Lippmann commented that German university students completed their practical education in factories, but people in industry only came to the German university for the theoretical knowledge indispensable to their careers; the influence of these people explained the great contemporary development in the German electrical industry. Darboux, the dean, added that German industrial progress was due to the students' freedom to learn what they wanted, unfettered by narrow regulations. Paul Janet, son of the philosopher of the same name and a specialist in industrial electricity, added that in France the license granted by the university was not welcomed in industry and brought suspicion on the holder rather than an advantage to him.[44] Clearly, some features of the interaction of industry and university in Germany could be considered for introduction into the French system, at least in the Sorbonne. Not everyone at the Sorbonne agreed with Lippmann and Janet, both of whom were deeply involved in certain technical aspects of science. But the industrially oriented group had become so numerous and powerful by the first decade of the twentieth century that it could change the direction of teaching.

The difference of opinion came out in a sharp exchange that followed the Sorbonne's introduction in 1907 of courses in industrial drawing and practical mechanics. A larger number of its science students were pursuing industrial careers but had a relative disadvantage because the Sorbonne lacked courses in these two areas. For a number of years the students had organized their own course in industrial drawing. The issue was examined by a faculty commission of Bouty, Janet, Koenigs, Moissan, and, after Moissan's death in 1907, Haller, the professors directly concerned. In spite of the strong case put forward by the Janet report, the dean emphasized that the faculty must not serve as a training ground for industrial schools. Picard seized this occasion to deliver a biting commentary on a trend he viewed as disastrous: ". . . L'université de

spécial (Paris, 1874), annexed to the _Procès-verbal, Conseil supérieur de l'instruction publique_, November 25, 1874, AN, F17 12959.

43. See Paul, "The Debate over the Decline of French Science," _French Historical Studies_, forthcoming.

44. _PV Sciences_, December 18, 1895, p. 37.

Paris n'a plus qu'un seul souci: fabriquer des produits pour l'industrie. Dans dix ans, si nous ne résistons pas aux injunctions des Etudiants, la Faculté ne sera plus qu'une Ecole d'arts et métiers. Nous devons penser d'abord à la Science pure et consacrer à son développement toutes nos ressources." Darboux added his familiar refrain: "Ce n'est pas en préparant à des écoles techniques que les facultés allemandes rendent la science utile à l'industrie." Bouty pointed out that 55 to 60 per cent of the physics students planned industrial careers and that the laboratories provided a good preparation. Janet's comment that not all the students could make their living as professors indicated the substantial increase in the number of students during the preceding few decades. Hadamard stated his belief that scientific theory benefited from experimental and even technical developments, but Picard replied that the technician represented the high point of empiricism and that he would certainly not change his teaching in the sense indicated by Hadamard. Even Haller thought that Janet's report placed too much emphasis on the technical fields. The Sorbonne must not desert pure science. The German chemical industry owed its prosperity to the resources it found in the universities. Even in the chemistry institutes, chemistry must be taught for itself. By refusing to give a common aim to the universities and the technical schools, France could avoid the sad antagonism existing in Germany.[45] Most of the faculty agreed on the need to introduce new courses but did not want to turn the Sorbonne into a higher Conservatoire.

In 1909 an interesting judgment on French science by a German was reported by Joseph de Moussac in an article on the German chemical industry. Moussac quoted with approval a long opinion of Robert E. Schmidt that generally resembled the opinions expressed by many French critics:

Vous avez en France des travailleurs remarquables, des chercheurs inlassables, des savants merveilleux. C'est là un fait indéniable et croyez-le bien, malgré leur chauvinisme, il n'est pas d'Allemand pour le contester. Malheureusement, chez vous ces sont des *isolés*, en ce sens qu'ils parviennent, dans leurs études, à faire une découverte de valeur, votre gouvernement leur décerne titres et honneurs, mais trop souvent ils n'arrivent pas à "monnayer" leur invention, c'est-à-dire à en trouver l'immédiate application dans l'industrie, et ils meurent pauvres.

45. Ibid., May 15, 1907, pp. 58–61.

En Allemagne, au contraire, dès qu'un ingénieur a fait la moindre découverte d'une application industrielle un peu pratique, il trouve aussitôt cent propositions de vente avantageuses. Aussi bien pour leur enlever jusqu'à la tentation de porter ailleurs leurs trouvailles, chacun de nos chimistes et de nos ingénieurs est directement intéressé à un prorata assez élevé dans tout ce qu'il produit comme invention ou même comme amélioration.[46]

Even Henry Le Châtelier, in some ways, like Janet, a high priest of industrial science, emphasized the care that must be taken in reorienting science at the Sorbonne. In 1908 a group of industrialists wanted to create a chair of applied chemistry for C. Chabrié. Le Châtelier observed that this move, coming after the Commercy bequest, was an interesting symptom of the times, although it could hardly be expected that French industrialists would do as much for education as their American counterparts. But the connections between industry and science were tightening. The case of Chabrié resembled those of Haller in chemistry at Nancy and Janet in electricity at Grenoble. Le Châtelier wanted to discuss the general ideas behind a technical education now that the first courses of an industrial character were to be introduced into the Sorbonne. Bouty, who had earlier wondered whether the nature of Le Châtelier's research qualified him for a Sorbonne appointment, now welcomed the development that brought industrialists closer to science, a trend also evident in the custom of a professor and an industrialist alternating as president of the Société des électriciens. Le Châtelier recommended a slow evolution of contemporary technical teaching that would reduce yearly the descriptions of apparatus and permit an increasing place for precise scientific ideas. ". . . C'est le rôle de l'université de tendre à donner à l'enseignement un caractère tous les jours plus scientifique." Darboux completely agreed with Le Châtelier and found his ideas compatible with those of Friedel, the "founder of applied chemistry."[47]

The trend toward practical courses sometimes occurred at the expense of basic science courses, but this trend was generally re-

46. M. le Comte Joseph de Moussac, "L'industrie chimique en Allemagne," *Revue des questions scientifiques* 66 (1909): 414–15.

47. *PV Sciences*, June 22, 1908, pp. 72–76. On Le Châtelier, see A. Silverman, "Henry Le Châtelier: 1850 to 1936," in Idhe and Klooster, *Selected Readings*, pp. 138–43, and F. Le Chatelier, *Henry Le Chatelier* (Paris, 1968).

sisted by the scientific establishment. In 1913 the University of Nancy wanted to replace its chair of geology with a chair of *électrotechnique*. Although both the Faculty of Sciences and the Council of the University at Nancy approved the change, the Section permanente of the Conseil supérieur de l'instruction publique rejected it. After a defense of mineralogy by some members of the section, Paul Appell, a specialist in rational mechanics who became dean of the Faculty of Sciences and then rector of the Sorbonne, warned of developing applied sciences at the expense of basic sciences. Since the applied sciences frequently received local subventions, it was even more important to maintain a minimum of chairs in the basic sciences. Darboux supported Appell: "L'essentiel actuellement est de ne pas sacrifier les chaires scientifiques aux chaires de sciences pratiques. Si elles sont mises sur le même pied cela ne pourra se faire qu'au détriment de la bonne entente dans la Faculté. . . . La minéralogie n'est pas une science inférieure, tant s'en faut. Il n'y a pas une seule Université allemande qui n'ait une chaire de minéralogie."[48] The excessively practical orientation of some of the provincial faculties had to be kept in check, especially in view of the municipal and departmental funds that were often available for courses directly related to agriculture and industry.

The war brought a definite sense of urgency to the plans for reforming the teaching of science in the Sorbonne. In 1915 the dean, Appell, spoke to the faculty on the necessity of preparing for the future a special program for training young people to qualify to serve industry and also to get to know scientific methods and laboratory practices. A commission of Le Châtelier, Lippmann, Koenigs, Janet, Gabriel Bertrand, and E. Vessiot was chosen to study the issue. The fact was that many who had acquired some scientific knowledge could not find any place to learn to work, and the heads of factories neglected laboratories because they did not know the services that could be obtained from them. The commission had to study the situation and evolve a plan; only in this way could France make the necessary adjustments and eliminate the advantages of Germany in this area: "Après la victoire il y aura à travailler encore ensemble à former des hommes capables de lutter dans le domaine industriel, le projet dont on vient de parler y aidera; car l'Allemagne ne renoncera pas de suite à ses positions acquises. Les jeunes gens instruits de France devront être également préparés à lutter sur ce

48. Session of June 13, 1913, AN, F17 13666.

terrain. Nous nous y emploierons tous."[49] A few months earlier Le Châtelier had declared that it would be interesting to have the German type of laboratory where one rapidly became familiar with methods currently employed in analytical chemistry. (Not much existed outside Nancy for practical instruction in analytical chemistry.) Le Châtelier specified that he wanted to see the means furnished for training the heads of experimental factories, not chemists, a large number of whom already existed.[50]

Political intrusion into scientific and technological questions became an issue during the war. In addition to Le Châtelier's plan for the creation of higher experimental education, Senator Goy put forward a plan to create faculties of technical sciences. A Sorbonne faculty commission examined the project and recommended the use of existing institutes rather than the creation of independent new entities. After the faculty had heard this 1916 report of Janet, Darboux warned against imitating foreign models in establishing technical establishments. France had been the pioneer in this area: "Les allemands n'ont fait que nous imiter. Ils ont la sagesse de maintenir le pont entre l'université et les Ecoles de sciences." He opposed creating technical schools within the faculty, which was a group of theoreticians. Rather, existing technical institutes should be developed with faculty support. Lippmann did not entirely agree; in Germany the Faculties of Science introduced themselves into industry. Picard agreed generally with Darboux. Janet pointed out that the commission's conclusions resembled the opinions of Darboux and Picard.[51] The faculty put forward its own counterplan to the Goy scheme: reforms were needed, but they should be carried out within the existing educational structure. One of the advantages in Germany, Le Châtelier pointed out, was that many chiefs of industry had simply done higher studies and were rewarded with a doctorate. This was needed in France: "Il faudrait en France beaucoup de tels chefs. Pour cela il faudrait créer un enseignement scientifique tout à fait élevé." The creation of technical institutes would not be enough. Picard saw no utility in technical faculties.[52] The de-

49. PV Sciences, February 26, 1915, pp. 144–45.
50. Ibid., November 14, 1914, pp. 194–95.
51. Ibid., February 10, 1916, pp. 149–51.
52. Ibid., March 2, 1916, p. 154. The Faculty of Sciences at Nancy was the only one out of six faculties to vote for the creation of "Facultés des sciences appliquées," on the grounds that they already existed in the institutes of the universities. The Sorbonne agreed to an industrial section for the Faculty of

bate had a certain futility about it because the momentum of the war pushed science in the direction of industry regardless of faculty opinion.[53] Enthusiastic government support for this trend made it difficult to resist. The faculty could only hope to ride the wave of enthusiasm for technology and control it in the interest of science.

As was made plain by one of the leading French scientists and the chief architect of wartime scientific and technological mobilization, the issue was national survival. In his speech before the public session of the Academy of Sciences in 1918, the president, Paul Painlevé, spoke of the role of science in the allied victory. He believed that the pursuit of science for purposes of immediate utilization of results, for pecuniary gain, or for oppression of peoples degrades the soul and ends in a sort of scientific barbarism: "La science n'est moralisatrice qu'à condition de garder aux yeux de l'élite qui la cultive son caractère essentiel qui est la recherche désintéressée de la vérité." This was the case in France, and the triumph of virtue in the war proved it. It was far different in Germany, argued Painlevé: ". . . de l'autre côté du Rhin, la Science, c'était une gigantesque entreprise où tout un peuple, avec une patiente servilité, s'acharnait à fabriquer la plus formidable machine à tuer qui ait jamais existé." In recognition of new trends, the Academy of Sciences in 1918 created a six-member division named "Applications de la Science à l'industrie." Painlevé praised the union of science and industry because of its remarkable results in the war effort and noted happily that the future of the union was safeguarded by the formation of the new academy section. Painlevé admitted that French scientific teaching perhaps dealt too much with general theories and ignored immediate applications, but at the same time it respected, stimulated, and developed individuality and the original and inventive faculties of the mind. These were the qualities that had saved France when faced by an enemy superior in nearly every respect: "Ce sont des qualités qui industrielle-

Science. "Documents et enquêtes," *Revue internationale de l'enseignement* 70 (1916): 53–68, 129–33, 280–308.

53. The Comité consultatif des arts et manufactures rejoiced in this development and predicted the course of the future: "L'état de guerre a crée la collaboration intime du gouvernement et de la science, de cette science qui devra se tourner sans délai vers l'industrie. Tout le monde s'est ressaisi et les merveilleux résultats obtenus, en quelques mois, dans nos usines de guerre, nous ont apporté la certitude de la victoire." Page proofs of the reports (1919) of the Comité are in AN, F12 8045.

ment ont sauvé la France; c'est parce qu'ils étaient doués d'imagination, de connaissances générales et de facultés créatrices que nos
ingénieurs ont pu faire face à une situation désespérée." Scientific
research, inventions, and national defense were organized systematically by the Direction des Inventions, created in 1915, which
was copied by the allies to the extent of setting up interallied relations. Since Painlevé was involved in the scientific mobilization
effort up to the end of 1916 and had great interest and considerable
influence in the politics of postwar international science, his opinion
indicated a new orientation in governmental scientific policy itself.[54]

54. Paul Painlevé, "Les raisons de la victoire des alliés et ses conséquences"
(discours prononcé par M. Paul Painlevé, Président de l'Académie des Sciences,
le 2 décembre 1918, à la séance publique), *Revue scientifique*, January 4–11,
1919, pp. 1–5. See ch. 4n23.

2. The Scientist as Propagandist

WHEN WAR WAS DECLARED in 1914[1] the German intellectual community sang a secular *Te Deum* to celebrate the rebirth of idealism, the end of the threat of materialism, and the death of politics implicit in the answer of the nation to the call of the Fatherland. Ecstatic outbursts embodying "the ideas of 1914" rivaled Wordsworth's ravings on the beginning of the French Revolution. Most of the German intellectual fury was directed against the English "trader." Much of what Sombart said against the bourgeois spirit and utilitarianism of English philosophy and science was an exaggeration of Pierre Duhem's famous analysis of English science in 1893. It was also the same complaint that the English and French hurled against German science in their wartime propaganda broadsides. No doubt Ringer observed correctly that this propaganda was chiefly for domestic consumption, an attack on all the domestic evils the German intellectuals had been denouncing before 1914: "The purpose . . . was to erect permanent symbols of the mandarins' own values and, if possible, to perpetuate the national consensus embodied in the 'ideas of 1914' beyond the period of the war itself." The delusion was equally strong among the French as a source of inspiration for pursuing the great massacre. However outrageous the war aims of the politicians, the demands of the large majority of intellectuals on both sides for a *Siegfrieden* made the politicians' aims appear to have been inspired by a fine sense of Aristotelian moderation. The German intellectuals' petition to the Imperial Chancellor in 1915 asking for extensive annexations contained, out of a total of 1,347 signatures, those of 352 professors. Among the scientists, Ostwald and Haeckel were orthodox in their extreme de-

1. For bibliography on philosophy and the war and science and the war, as well as a host of other topics, see *The New York Public Library Subject Catalog of the World War I Collection,* 4 vols. (Boston, 1961).

mands. Another petition containing more moderate demands got only 141 signatures, 80 of them from professors.[2] Chomsky's American mandarins are in the grand tradition.

The reaction of the French intelligentsia to the war was just as enthusiastic as that of the German mandarins. Not all were young enough, like Péguy, to indulge their death wish on the battlefield, but all could beat their pens into swords. Henri Bergson was in the forefront. In his presidential address to the Académie des sciences morales et politiques on December 12, 1914, he denounced Germany as a mechanism: "Throughout the course of the history of Germany . . . there is, as it were, the continuous clang of militarism and industrialism, of machinery and mechanism, of debased moral materialism."[3] Emile Boutroux was more rational in his approach to the

2. Fritz K. Ringer, *The Decline of the German Mandarins. The German Academic Community, 1890–1933* (Cambridge, Mass., 1969), especially pp. 180–99. See also Klaus Schwabe, "Zur politischen Haltung den deutschen Professoren im ersten Weltkrieg," *Historische Zeitschrift* 193 (December 1961): 601–34; Fritz Fischer, *Germany's Aims in the First World War* (New York, 1967). The first petition may be found in Salomon Grumbach, *Germany's Annexationist Aims* (London, 1917). See also *Out of Their Own Mouths* (New York, 1917). An important collection of documents on the propaganda of the intellectuals may be found in the *Revue scientifique*, August 8–November 14, 1914, pp. 170–76 ("Notes et actualités"):

    I: L'appel des intellectuels allemands aux nations civilisées.
    II: Quelques appréciations sur la guerre et quelques réponses aux
         manifestes des intellectuels allemands.
        1. Académie des sciences.   2. Académie française.
        3. Intellectuels anglais.   4. Intellectuels russes.
        5. Paul Appel, Président de l'Institut de France.
        6. William Ramsey.
        7. Charles W. Eliot, Président de l'université Harvard. Etc.

See ibid., September 30–October 7, 1916, pp. 599–600 ("Variétés") for "Les responsabilités de la guerre—Message des intellectuels américains aux peuples des nations alliées. 500 notabilités du monde des lettres, des sciences, des arts, de la politique, de la finance, de l'industrie, et du commrce [*sic*]." *Le Temps*, November 26, 1914, presented Ostwald as one of the most effective German propagandists in Sweden. Haeckel's prowess as propagandist may be seen in *Das monistiche Jahrhundert*, November 16, 1914, nos. 31–32. See also M. Caullery, "Ernst Haeckel et son 'évolution' à propos du militarisme," *Revue scientifique*, November 11–18, 1916, pp. 673–78. Caullery, who taught biology at the Sorbonne, was the first French exchange professor in the sciences at Harvard. Caullery noted that the biologist Richard Hertwig resigned from the Société zoologique de France.
3. Henry Bergson, *The Meaning of the War: Life and Matter in Conflict* (London, 1915). This translation went through three impressions from June to August, 1915.

war, as he had been in his philosophy, but he still saw the three principles of German culture as force, organization, and science— "man is literally reduced to the condition of a machine."[4] All the learned journals and periodicals teemed with articles by those intellectuals and academics who knew anything about Germany.[5] Frequently, a former advocate of emulation now proved his patriotism by leading the chorus denouncing the Teutonic Moloch.

One French response to the pronunciamentos of German intellectuals was to solicit opinions of leading French intellectuals, especially scientists, on German science, widely regarded as the chief area of challenge. Organized and edited by Maurice Leudet of *Le Figaro* and Professor Gabriel Petit, the opinions were published in the newspaper, beginning in April 1915, and as a book in 1916.[6] A short preface set forth the salient points of German claims that most alarmed the French: the brain of Europe could not be other than that of Germany; the German empire will bestow a perfect civilization upon humanity; Germans are incomparably superior, morally and intellectually, to all others; the Germans are the chosen people of the earth. To Petit and Leudet a German *Appeal to Civilized Nations*, signed by ninety-three German intellectuals, embodied the formulae of a mad pan-Germanism showing how the German soul had remained barbaric under its insolent mask of *Kultur*.[7] The

4. Emile Boutroux, *Philosophy & War* (London, 1916). This volume contains most of what Boutroux said about the war. See "L'Allemagne et la Guerre," *Revue des deux mondes* 23 (October 15, 1914): 385–401.
5. The *Revue de Paris* for 1915 carried articles by Gustave Lanson, "Culture allemande, humanité russe," December 15, 1915, pp. 333–48; Ernest Lavisse, "Trois idées allemandes," May 15, 1915, pp. 225–35; and Charles Andler, "La doctrine allemande de la Guerre," January 15, 1915, pp. 263–84.
6. Gabriel Petit et Maurice Leudet, *Les allemands et la science* (Paris, 1916), hereafter cited as Petit. Much used by Charles Nordmann, "A propos de la science allemande," *Revue des deux mondes* 31 (January 15, 1916): 457–68, a sequel to his "Science et guerre," *Revue des deux mondes* 30 (December 1, 1915): 698–708. As a result of the Panama Pacific International Exposition at San Francisco in 1915, a useful compilation of French work in all scholarly areas appeared, with a bibliography of the chief works, no doubt also encouraged by the war and the threat from Teutonic *Kultur*: *La science française*, 2 vols. (Paris, 1915). A completely revised edition was published in 1933. The surveys were by experts in the various areas.
7. A great deal of the debate over the relative virtues and vices of German and French science is obviously related to and often part of the sticky debate over national character. In *Les allemands et la science*, the sections by Maurice Barrès, Marcellin Boule, A. Dastre, Emile Boutroux, E. Gautier, and Pierre Duhem contain observations on German national character. See H. C. J. Duijker and N. H. Frijda, *National Character and National Stereotypes* (Am-

editors recognized that their collection contained a great deal of irony and passion but hastened to point out that it remained an elegant and timely commentary on the German aim of scientific hegemony: "In response to so many perfidious insinuations from across the Rhine, which should have incited us to react well before the war it [the book] shows decisively that France, far from declining, has never ceased to be . . . an incomparable initiator."[8]

The general tone of the collection was set in a preface by Paul Deschanel, president of the Chamber of Deputies. Drawing upon the polemical conclusions of the contributors, Deschanel gave proof of the qualities that had put him in the company of the Immortals by summing up in a few concise and elegant paragraphs the case against the claims of the *Doctus Bochensis*.[9] With certain exceptions, the Germans had excelled in developing discoveries made by others. As Deschanel pointed out, the essays were a striking commentary on the declaration of the Académie des Sciences on November 3, 1914, part of a general protest of the other academies of the Institute: for three centuries most of the great discoveries in the mathematical, physical, and natural sciences, including the chief inventions of the nineteenth century, were products of the Latin and Anglo-Saxon civilizations. For the German, Deschanel argued, science, history, philosophy, and religion were national forces like the army, diplomacy, and credit. No longer universal and human, science was first at the service of the state. If this were so, it could well have been a lesson learned, like so many others, from the French, whose marshaling of the regiments of Isis in the Revolutionary and Napoleonic periods had established a trend that would reach its apogee in the mid-twentieth century.[10]

One of the motifs of the essays, Deschanel noted, was the idea that French and German contributions to science had been entirely

sterdam, 1969). Its bibliography is Germanic. Another dimension of the debate is revealed in the scorn of Ostwald and Haeckel for decadent "Latin" civilizations; this had repercussions in Spain. José Aubin Rieu-Vernet, ¿La inferioridad de la ciencia francesa? (Madrid, 1918), defended the superiority of the French achievement in science, thus indirectly denying "la inferioridad de la raza y de la ciencia latinas."
8. Petit, pp. xix–xx.
9. Yves Delage, "Histoire naturelle du DOCTUS BOCHENSIS (olim germanicus, obsol.). Variété mal connue de l'espèce Homo sapiens (Linn.). Anatomie, Physiologie, Mœurs, Industrie," ibid., pp. 99–115.
10. See Maurice Crosland, The Society of Arcueil. A View of French Science at the Time of Napoleon I (Cambridge, Mass., 1967).

different in nature. It is necessary to distinguish between invention and genius, on the one hand, and the works produced by the university, industrial, and commercial application of seminal ideas, on the other. France wore the laurels of the creator in science: Descartes and Pascal in the seventeenth century, Lavoisier in the eighteenth century, and Pasteur in the nineteenth century. This Procrustean proof was supported by an enumeration of the glorious creators in French science in the first half of the nineteenth century.[11] With a perceptible nostalgia, Deschanel looked back to the seminal days of Cauchy, Ampère, Berthollet, and Lamarck, when "all nations came to school in France." Was the teacher soon surpassed? Certainly not, Deschanel argued, with the continuation of Poincaré, Hermite, Bertrand, Fresnel, Sadi Carnot, Becquerel, and Curie. Against the great German chemists Liebig, Bunsen, and Kekulé, one could easily set Dumas, Wurtz, Sainte-Claire Deville, and Pasteur. Other areas could be matched with equal ease. The impartial judgment of the contributors to Petit and Leudet's anthology would show that the German claim of conducting the scientific "orchestra" was quite unjustified, as was the Wagnerian arrogance with which they were trying to drown out other more important members of the group.[12]

An extraordinarily harsh deflation of German claims came from the pen of Emile Picard. It was not surprising that he would put forward Frenchmen like Cauchy, Fourier, and Galois as pioneers who had directed modern research toward pure mathematics and mathematical physics and that he would see celestial mechanics since Newton as a French science (d'Alembert, Lagrange, Laplace, Poincaré). In astronomy and in the foundation of modern chemistry, the English and French took full honors. In thermochemistry and physical chemistry the Americans and the French (Berthollet, Sainte-Claire Deville, Berthelot) were advanced as the key figures. Credit for the first law of thermodynamics was given to Marc Séguin, not Mayer. Besides, why mention anyone after Sadi Car-

11. The list was quoted from Gaston Darboux, secrétaire perpétuel de l'Académie des Sciences, as found in H. Laurens, ed., *L'Institut de France* (Paris, 1907). There was, of course, an incomparable galaxy from which to choose, just as there was in Germany later.

12. For the best summary of the factors in the "decline" of French science, see Robert Gilpin, *France in the Age of the Scientific State* (Princeton, 1968). For criticism of Gilpin and of others who have dealt with this issue, see my forthcoming essay in *French Historical Studies*.

not?[13] Röntgen got a grudging admission to the group. Picard concluded that the Germans had, throughout the ages, rarely shown any great originality, although they were now showing skill in exploiting the ideas of others, which the industrial character of their civilization permitted. (*"Mais il ne faut pas confondre l'augmentation du rendement scientifique et le progrès réel de la science."*) His brief history of the sciences showed the bizarreness of the German intellectuals' proclamation that the Germanic race alone was capable of working for the development of *Kultur*.[14]

Several of the twenty-eight contributors, especially E. Gley, medical scientist and professor at the Collège de France, pointedly excluded science from the possibility of contamination by nationalism. Literature, art, and even philosophy could express the genius of a race, represent the spirit of a nation, and embody the soul of a country, but this was not possible for science.[15] Gley transferred the theological concept of the immutability of dogma from the religious to the scientific sphere. Gley's statement could have been subscribed to by many of the scientists and physicians in republican favor. The scientific philosophies of people like Duhem and Poincaré could not yet be incorporated into a viable scientific ideology that

13. On the issue of priority in scientific discovery, see Robert K. Merton, "Priorities in Scientific Discovery: A Chapter in the Sociology of Science," *American Sociological Review* 22 (1957): 635–59, reprinted in Bernard Barber and Walter Hirsch, eds., *The Sociology of Science* (New York, 1962), pp. 447–85; and Merton, "Behavior Patterns of Scientists," *The American Scholar* 38, no. 2 (1969): 197–225.

14. Emile Picard, "L'Histoire des sciences et les prétentions allemandes," in Petit, pp. 284–89; *Revue des deux mondes* 28 (July 1, 1915): 55–79; and *L'histoire des sciences et les prétentions de la science allemande*, 2d ed. (Paris, 1916), published in the series "Pour la vérité. Etudes publiées sous le patronage des secrétaires perpétuels des cinq académies." Paul Bert's successor in physiology at the Sorbonne, A. Dastre, in "Du role restreint de l'Allemagne dans le progrès des sciences," in Petit, pp. 84–96, gives another, slightly more detailed summary showing French pre-eminence in science and German confusion of organization with science itself. The same theme is found in Sir William Ramsey, "La part médiocre des Germains dans la découverte scientifique," in Petit, pp. 325–33, a survey that made the collection of essays technically a joint Anglo-French effort. It is amusing to find Sir William using Ludwig Darmstädter and Emil du Bois-Reymond, *4000 Jahre Pionier-Arbeit in den exakten Wissenschaften* (Berlin, 1904), for names and statistics, although he used them against the Germans. But then du Bois-Reymond, as Sir William noted, was of French extraction! See also Rieu-Vernet, pp. 238–40: "Apéndice—Declaraciones de un sabio catedrático de medicina español [Juan Madinaveitia]: Los sabios alemanes se apropian los descubrimentios ajenos."

15. E. Gley, "Science et savants," in Petit, p. 180.

was politically acceptable. Only after the Republic sensed itself the victor in the struggle with the church could it permit a tampering with the immutable base of the ideology that had sustained it in battle.[16] Gley proclaimed his faith in a theory on the development of science that was, to some degree at least, an intellectual anachronism by 1914: scientific works are founded on what is permanent and less individualistic in the human mind; they are supported only by reason in the attaining of their unchanging goal, the knowledge of truth. In a structuralistic vein, Gley declared that truth and reason are universal and that the laws of logical thought are the same everywhere at all times. To speak of a national science would be to accept a proposition contradicting the very idea of science. The facts of science and the natural laws embodied in their necessary and mutual relations were the same in Germany as in France. Science escapes all environmental influences. Another point was more solid: science is a collective work, especially in the age of rapid transmission of scientific information. Since science is an impersonal, universal, and collective work, it is an absurdity to speak of German or French science.[17]

Another medical scientist, Joseph Grasset, recently retired from Montpellier, refused to write about German science: "In spite of the war and the atrocities of the barbarians who have attacked us, I remain anchored to my old bourgeois idea that *there is not a German science and a French science;* there are German scientists and French scientists."[18] Although methods of procedure can be compared, science is an interacting totality that cannot be judged in separate national components, "whatever may be the respective value of the . . . stones each brought for the construction of the edifice."[19] Grasset was, perhaps, keenly aware that the same criticisms could be leveled against French scientists—indeed, the Petit project gave eloquent proof of this. German scientists could be criticized for only one thing: they gave their opinions as German citizens but made it appear that these opinions were based on sci-

16. See Henry E. Guerlac, "Science and French National Strength," in Edward Mead Earle, ed., *Modern France* (Princeton, 1951), p. 89, and Paul, "The Debate Over the Bankruptcy of Science in 1895," *French Historical Studies* 5, no. 3 (Spring 1968): 299–327.

17. Gley, in Petit, pp. 180–83. For a recent view of science as a world-wide system little affected by local policies, see Derek J. de Solla Price, "The Science of Scientists," *Medical Opinion Review* 10 (1966): 88–97.

18. Dr. Grasset, "Lettre au professeur Gabriel Petit," in Petit, p. 200.

19. Ibid., p. 201.

entific authority. The German scientists had subscribed to immoral and senseless declarations, but their scientific work lent no authority to the acts, for they were acts of German citizens. Science was not responsible. But few could separate science and patriotism, as this famous Catholic conservative medical scientist attempted to do.[20]

The Nobel prize winning physiologist and pacifist Charles Richet, professor in the Faculty of Medicine at Paris, shared, to some extent, Grasset's opinion that science could not be treated in terms of national characteristics. Richet condemned the prevalent propaganda that "French science is everything; German science is nothing," and emphasized that he did not intend to let his patriotism influence his appreciation of German scientific work. Certain activities of the mind can be classified as national: French prose, Italian comedy, and German poetry had exact meanings because the terms meant that the prose, comedy, and poetry were written in the designated languages. But science, like sculpture, music, and numismatics, cannot be described as belonging to a particular country. Richet assigned great discoveries to individual men rather than to abstract entities like nations. He had a strong penchant for seeing science in terms of personalities and individual "break-throughs." (The age was several decades away when it would be common to use such concepts as independent multiple discoveries or to see Germany and the United States as countries where decentralization of the educational systems and an accompanying competition produced world leadership in science.) The facts that Huyghens and Leibniz wrote in French and Latin and Schwann lived in Belgium showed, Richet argued, how difficult it was to assign a nationality to a discovery. But he followed the approach of Gley rather than that of Grasset: one could attempt to appreciate the contributions of German and French scientists, as Richet did for the biological sciences.[21]

Only Boutroux, who probably knew German intellectual life better than most of the contributors, limited as they were to some solid knowledge of their specialties, stated that German science was the most authoritative in the world. In spite of the petty criticisms of its lacunae and imperfections—that the Germans excelled in the

20. On Grasset's politics, see Paul, "The Crucifix and the Crucible: Scientists, Religion, and Politics in the Third Republic," *The Catholic Historical Review*, forthcoming.

21. Charles Richet, "Science française et science allemande," in Petit, pp. 346–49.

mechanical parts of scientific work rather than in invention, that their exposition was confused and obscure, and that practical applications were dominating disinterested research—the prestige of German science seemed indestructible. Germany was the born teacher of the world. *"Wir sind die Meister aller Welt / In allen ernsten Dingen"* was the popular statement of this idea in an anthology of verse and songs compiled for German soldiers in 1914. But Boutroux decided to take a new look at the case in favor of German science.[22]

Boutroux conceded that one must justly recognize that German science adhered to the religion of competence. But what was this competence? It was natural that the author of works in the philosophy of science[23] would ask whether a man was really competent if, on the basis of a preconceived idea, he eliminated from scientific research all living, human, personal, and rational elements in order to depend only on materially objective data and on a process of reasoning that excluded all intuition.[24] The critical point in German science was the transition from fact to idea, which in great thinkers like Galileo, Newton, and Descartes was made by a continuous movement of the intelligence that progressively disengaged the general from the particular. In the scientific determination of facts, the mind had already come into operation, and, in supporting itself constantly on facts, it reached the highest ideas. Method is constant contact with facts during incessant activity of the intelligence. The Germans, thinking such a method too subjective and too human, began to look for ideas in the transcendental world, without connection with the world of facts. But Boutroux' example came from Schelling, not Euler: "C'est ainsi que le philosophe Schelling en vint à déduire, de l'identité primordiale A = A, la loi newtonienne de l'attraction ainsi que la dualité des fluides électriques, et à corriger la nature, là où celle-ci s'avisait de lui désobéir."[25] When the method had to be abandoned, German science replaced it with a procedure in which there was a pure and simple identification of the idea with the totality of facts in a given category. In the area of the practical, the German scientist believed that he alone possessed the totality of facts and was alone capable of determining

22. Emile Boutroux, "La science allemande," in Petit, pp. 48–49.
23. *De la contingence des lois de la nature* (Paris, 1874); *De l'idée de loi naturelle dans la science et la philosophie contemporaine* (Paris, 1895).
24. Boutroux, in Petit, p. 52.
25. Ibid., p. 53.

the guiding ideas of science. Boutroux argued that the totality of facts in any order of things is a chimera, and the German scientist really arranged facts as it seemed proper to him; then starting with his generalizations as incontestable axioms, he revealed to the world the meaning of particular events while guided by the needs of his case. There was no possibility of questioning the German scientist since he believed he knew each fact best and was interpreting it as a function of the total pattern. Cartesian "good sense" could not be invoked because that would introduce a subjective element. Molière had given the best judgment of such a method, a judgment that mankind would soon echo: "Raisonner est l'emploi de toute ma maison, / Et le raisonnement en bannit la raison."[26]

Boutroux' argument was partially accepted by the paleontologist Marcellin Boule, professor at the Muséum national d'histoire naturelle. He willingly admitted that the most imposing facet of German science was its method of organization. On the one hand, it functioned like an immense factory, perfectly equipped, with an excessive output, quantitatively if not qualitatively. On the other hand, it presented the aspect of perfect functioning in the domain of applied science. From any new principle or great theoretical discovery, it could draw all the principles of a material order that would increase German political, military, and industrial power. Such a science was a means rather than an end and lacked "noblesse." Although German science was of inferior quality, involving obscure preliminary work, weighty details, copious memoirs full of useless things—in short, monstrous scaffolds, not monuments—it was aped by other nations. Boule thought this strange and commented on the paradox. Instead of coming to the inconvenient conclusion that he had caricatured a minor aspect of German science and that it was widely imitated because some of the major aspects of that science were Faustian, Boule saved his image of German science by the stock argument focusing on the defeat of 1870–71. After their defeat the French aped the work methods and gorged themselves on the thoughts of their conquerors. Intoxicated, the germanified masters carried out a program of germanification of

26. Molière, *Les femmes savantes*, act 2, scene 7, Chrysale to Belise. Her reply could serve as a German reply to the French and especially as a German opinion on English science:

Est-il de petits corps un plus lourd assemblage,
Un esprit composé d'atomes plus bourgeois?

their pupils' minds. This is obviously an exaggeration, for the imitation of Germany in many areas antedates 1870, and the defeat actually brought about a reversal of the process in some areas. What Boule thought began in 1914 may be found, for the most part, in a considerable number of fields after 1870.[27] In certain spheres, like the building of laboratories, Pasteur and others had called for emulation of Germany. But Boule wondered how Frenchmen could have failed to see that the superiority of German science was a peripheral superiority of a material sort, lacking genius, and leaving to others the task of brilliant syntheses. Boule was one of many Frenchmen whose nearly pathological cult of clarity, along with their retention of the Enlightenment syndrome of French superiority, made them fail to see or at least caused them to underestimate the scientific "revolutions" developing outside France in the late nineteenth and early twentieth centuries. Boule proudly recalled that he had always opposed the "snobisme" of germanophilia, although he was a *vox clamantis*, able to do little against the professors of the university and the authority of the great French thurifers of German science. The events of 1914 would enable France to regain her sanity!

But Germany's power had been used exclusively to dominate and oppress other peoples, and this type of hypertrophy had been accompanied by a corresponding atrophy of other aspects of the intelligence. Concentration on the utilitarian features of science had led to the near annihilation of the more noble factors leading to the Ideal, especially the pursuit of disinterested science, which alone leads to truth, justice, and the feeling of moral beauty. While attaining material progress, Germany had regressed spiritually and morally to the point where it deserved to be ostracized from truly civilized nations; the German cult of the baser aspects of science had reduced its people to the level of a primitive *Homo ferus*. Drawing upon Edgar Quinet's "scientific" historical laws, as did another writer in the collection, Boule concluded with an argument from biology and paleontology (the "biology of the past"), that Germany would follow the creatures shown by science to have fallen victim to their own gigantism or their exaggerated specialization in one direction only. So *Kultur*, the victim of its own materialist gigantism, would become extinct, like the reptile *Diplodocus* and

27. See Claude Digeon, *La crise allemande de la pensée française 1870–1914* (Paris, 1959), which does not, however, deal with science.

many other denizens of Secondary and Tertiary times. Obviously the French would escape this fate. Their conservation of the equilibrium of their faculties and of individual initiative permitted them to adapt to new conditions of existence according to the natural laws of "our great Lamarck." It would be another triumph of mind over matter.[28]

Several contributors selected as one of the vitiating characteristics of German science a factor that some modern commentators view as a key factor in its success: organization.[29] Boutroux saw organization as the general characteristic of German scientific work. In order to achieve the two qualities that made research valuable—*Vollständigkeit* and *Gründlichkeit*—and which were beyond the capacity of the individual, it had been necessary to organize teams where each man pursued his special function. It was absurd to deny the value of such an organization in examining all the aspects of a scientific issue and in giving science a solid, vast, and practical character that could not be surpassed. But Germany seemed to think that organization was all there was to scientific endeavor. This approach tended to shackle rather than favor intellectual activity, which remained, in Boutroux' opinion, the most important condition for scientific research. Was there not implicit in the German method the thesis that ideas in the guise of laws of nature emerge from assembled and coordinated data by a sort of spontaneous generation? Boutroux argued, however, that the history of science showed that human intelligence is the source of scientific ideas, which are, as Claude Bernard said, hypotheses going beyond the intrinsic significance of the raw data.[30] But is it possible to establish conditions favoring intellectual fertility? Yes, answered Boutroux, through an education of the mind that develops a sense of reality, a faculty of generalizing without escaping from reality. This education requires the scientist to leave the narrow limits of specialization and associate with workers in other specialities, but men who dominate their specialities, who think as men while work-

28. Marcellin Boule, "La guerre et la paléontologie," in Petit, pp. 34–45.
29. See, e.g., Joseph Ben-David, "Scientific Productivity and Academic Organization in Nineteenth-Century Medicine," *American Sociological Review* 25 (1960): 828–43, reprinted in Barber and Hirsch, eds., *The Sociology of Science*, pp. 305–28.
30. See Claude Bernard, *An Introduction to the Study of Experimental Medicine*, trans. Henry Copley Greene (New York, 1926, 1961), especially Part One, "Experimental Reasoning."

ing as specialists. Science must be guided by Descartes' "bon sens," the link between thought and reality, the true source of discovery and judgment. It is clear that Boutroux thought that German science had deviated from the path of virtue set down by the patron saint of French science, with sad results for itself and for humanity.[31]

Much the same opinion on the organization of German science was held by the physiologist and biologist A. Chauveau. In concluding a short section dramatically advertising the later theme of Polanyi, that science thrives on liberty,[32] Chauveau admitted that Germany had gained undeniable advantage over her rivals because of the systematic installation of laboratories and institutes, which greatly favored the teaching of pure science and especially the extension of practical applications. But Chauveau denied that this development had produced any progress in pure science itself: the discovery of principles and laws comes exclusively from the spirit of invention, an innate aptitude that no disciplined organization is able to create entirely, which always makes evident its productivity, even outside the help given by such organization.[33] This judgment was restated half a century later with scholarly sobriety and applied to the entire European scene by Kaplan: "The Growth of the *Herr Geheimrat* tradition of hierarchical authority in the German universities and the stagnation which set in after a period of vigorous growth during the middle of the nineteenth century had already set the seeds for the subsequent decline of research in the European universities which was not fully recognized until the late 1950s." But Kaplan argues that "properly reconceived, organizations and scientists can contribute much to each other's growth and development."[34]

31. Boutroux, in Petit, pp. 49–51.

32. Michael Polanyi, "The Republic of Science. Its Political and Economic Theory," *Minerva* 1, no. 1 (1962): 54–73. For a different view, see Amitai Etzioni, "On the National Guidance of Science," *Administrative Science Quarterly* 10 (1966): 466–87, and *The Moon-Doggle. Domestic and International Implications of the Space Race* (New York, 1964), especially chapter 3. For an overview of the issue, see Harvey Brooks, *The Government of Science* (Cambridge, Mass., 1968), especially chapters 1 and 3.

33. A. Chauveau, "La science vit de liberté," in Petit, pp. 70–74. The same hostility to organization as the key factor in promoting the progress of science is evident in the exaggerations of Dastre, in Petit, pp. 84–85, 96–97.

34. Norman Kaplan, "The Western European Scientific Establishment in Transition," *The American Behavioral Scientist* 6, no. 4 (1962): 17–21, and "Organization: Will it Choke or Promote the Growth of Science?" in Karl B. Hill, ed., *The Management of Scientists* (Boston, 1964), pp. 103–27. But Philip

A common complaint against German science was its verbosity and obscurity. Although Gley thought there was no such thing as German science, he recognized that there were German scientists and that it was a psychological necessity that they be affected by national influences, such as climate of opinion and education: "The *matter* of science . . . is always the same everywhere; what can differ according to nations is the *form* of science, the form in which the results of scientific research is [*sic*] presented . . . the expression given to it."[35] Even if scientific research were done in all countries according to the same method, each people had a style for stating the results. The French style was preferable to the German: the tediousness and the illogicality of the German language frequently made presentations verbose and obscure.[36] These defects of the language were sometimes aggravated by the defects of the German mind: tediousness, abuse of abstraction, and lack of criticism. Of course, there were German scientists, like Felix Hoppe-Seyler, Gustav von Bunge, and Otto von Fürth, who wrote with precision and sufficient clarity, but they, along with some others, were exceptions. Gley warned that one should not exaggerate the weak-

---

H. Abelson, "German Technological Resurgence," *Science* 165, no. 3891 (1969): 339, concluded that "the Germans . . . by temperament . . . particularly suited to high technology . . . are moving confidently toward becoming No. 1 in science and technology."

35. Gley, in Petit, p. 184.

36. F. Henneguy, "L'Allemagne et les sciences biologiques," in Petit, pp. 206–17, provided an amusing analysis of a typical German memorandum, which could usually make the essential point in one-tenth the length. See also Picard, ibid., pp. 290–91, who thought great scientists like Gauss and Weierstrauss often incapable of the same luminous clarity as Lagrange and J.-B. Dumas. The reader who wishes to pursue this topic may begin with Karl Vossler, *The Spirit of Language in Civilization* (London, 1932), chapter 7, "Language and Science"; and R. Priebsch and W. E. Collinson, *The German Language* (New York, 1938), chapter 9, "The Genius of the German Language." Alfred Malblanc, *Stylistique comparée du français et de l'allemand*, 2d ed. (Paris, 1963), p. 291, gives a suitably Solomon-like judgment: "Nos deux langues de culture apparaissent finalement toutes deux par des caractères différents comme des avancées vers la science et vers l'art. Le français appelle la science par sa prédilection pour le jugement et son sens de la causalité, sa reconstruction du réel, il prédispose à l'art par son sens de la métaphore et du symbole, de la figure; l'allemand appelle la science par sa précision, son amour du détail, sa GRÜNDLICHKEIT et prédispose à l'art par son heureuse transcription du mouvement et ses images pittoresques; dans les deux langues des qualités distinctes, mais complémentaires et fécondes." Those who wish to make some comparison may peruse Hans Fromherz and Alexander King, *Französische und deutsche chemische Fachausdrücke. Ein Leitfaden des Chimie in französischer und deutscher Sprache* (Weinheim, 1969).

nesses of these German works, for they were weaknesses of form, not of matter; the foundation remained the same, even if the exposition were often defective.[37]

Yet Gley thought the negative part of his analysis to be more important than the positive because the defects of form were intrinsic vices related to the nature of the Germanic mind. He then gave a description of German science similar to Boutroux' model. "It is the idea connected with the discovered fact that really constitutes the discovery,"[38] not the fact itself. Scientific knowledge, Gley argued, comes from the study of the permanent relations between phenomena; the real relations between the facts form natural laws. He divided the process into two stages: the purely experimental step of discovering the fact; and the experimental and rational step of searching for the constant relations between the facts, possibly including the determination of laws deduced from these relations, a step which would require new investigations and comparative experiments. This allows a great role for discursive thought. Beyond this activity lies the possibility of the creation of a scientific theory. Once beyond the first step, reasoning plays a large role, and this purely mental instrument, Gley argued, was not the same for all groups but differed according to men and race.

Sensing that he had perhaps moved away from his original proposition about permanent aim and invariable method in science, Gley recovered by pointing out that these remarks applied only to ideal science, not to science as actually practiced by individuals! He used Duhem's familiar arguments to prove his case.[39] Dominated by the deductive spirit, German scientists lacked critical sense, abused "systematizations," and indulged in metaphysical speculation. In summary, the absence of a critical outlook led to the establishment of natural laws on the basis of relations insufficiently studied, without the necessary control experiments having been carried out. Unlike the French, German scientists could not construct scientific theories that were real and rational at the same time. The national characteristics of Germany vitiated its science. Only the great masters of German science managed to escape the vices inherent in the German mind; their science had no national

37. See also Gautier, in Petit, pp. 168–77.
38. Claude Bernard, *Introduction à l'étude de la médecine expérimentale* (Paris, 1865), p. 61, quoted by Gley, in Petit, p. 187.
39. Pierre Duhem, "Quelques réflexions sur la science allemande," *Revue des deux mondes* 25 (February 1, 1915): 657–86.

character.[40] Like many of his fellow critics, Gley was hardly consistent: he criticized German scientists for developing scientific theories that were remote from reality as well as for amassing facts without any concern for synthesizing them into a theory. Except for Boutroux and Duhem, and possibly one or two others, the French critics were rather confused on the nature of scientific theory, especially the relationship between "fact" and "theory."[41]

Faced with the overwhelming quantitative superiority of Germans in scientific literature, the Frenchman who did not wish to admit the scientific leadership of Germany usually selected another equally nebulous criteria for judging scientific leadership, the quality of production. This was the tactic of Richet in his judgment on the comparative contributions of French and German scientists. To judge scientific value by number rather than quality of scientific works is relatively easy, but such statistics could lead to serious mistakes. The number of Scandinavian, Russian, and Dutch scientists writing in German and of Spanish and Italian scientists writing in French was large enough to make it impossible to make any accurate estimation of national contributions on the basis of the language used by the scientist. Richet found that for the period 1850–1910 the percentages of publications in physiology in various languages were German 48 per cent, French 30 per cent, English 12 per cent, Italian 6 per cent, and other languages 4 per cent. In surveying medical science as a whole for the year 1907, he arrived at these figures for publications in various languages: German 37 per cent, French 26 per cent, English 25 per cent, Italian 5 per cent, and other languages 6 per cent. The jump in English was due chiefly to the increase in publications in the United States. The rough figures for biological and medical publications were 45 per cent for the German language and 30 per cent for French, figures that also gave the quantitative relation of the German language population to that of the French. This meant, surprisingly, that the proportion of medical and biological publications was about the same for each one thousand inhabitants in both Germany and France. But Richet did not attach much importance to the statistical legerdemain he had performed. Thousands of small insignificant and inaccurate notes meant very little. The important thing was to

40. Gley, in Petit, pp. 183–90, 197–98.
41. See Thomas S. Kuhn, *The Structure of Scientific Revolutions* (Chicago, 1962).

know what new avenues had been opened in the biological sciences by the publications.[42]

In Richet's value structure, two names towered above the rest as great creators, Lavoisier and Pasteur. Richet regarded Lavoisier as having established the sacrosanct basis of physiology as well as of chemistry. Richet quoted the fructifying principle of Lavoisier: *"La vie est une flamme; les êtres vivants sont des êtres à fonction chimique. La combustion vitale est la source de la force."*[43] The advent of Pasteur brought the renovation of therapeutics, hygiene, and all the medical sciences: *"Les maladies sont dues à des parasites: et sans parasitisme il n'est que de rares maladies"* was the great principle Richet quoted to sum up Pasteur's innovations. Darwin, "who knew how to develop the idea of our great Lamarck" and whose ideas dominated the natural sciences, was placed only a little below this divine dyad.[44] But obviously other scientists, many of them German, had produced excellent works in the nineteenth century. In a hierarchical list he recognized as being absurd and arbitrary, Richet enumerated four Frenchmen, three Germans, and one Englishman: Lamarck, Bichat, Jean Müller, Magendie, Schwann, Claude Bernard, Lister, and Helmholtz. He recognized that it was difficult in most cases to establish any national pre-eminence. In electricity one could easily count four Englishmen, three Italians, and two Germans as initiators, and physiology was a similar case. Although he thought it possible to argue that it often seemed a characteristic of French genius to make primary discoveries in various areas, Richet admitted it was really impossible to establish the pre-eminence of any national group in the important work following

42. Readers inclined to accept Ben-David's arguments in favor of German and American superiority in the medical sciences should perhaps use Richet's useful corrective, although it opens up the possibility of an endless debate about significance and quality.

43. Richet, in Petit, p. 351.

44. Ibid., p. 352. Michael T. Ghiselin, *The Triumph of the Darwinian Method* (Berkeley, 1969), asserts that the work of Lamarck has been sterile in furnishing working hypotheses for biological research, whereas Darwin's work has been the chief source. This may be an "Anglo-Saxon" view. French biologists have always been eulogistic of "our great Lamarck." Most modern French biologists would probably agree with Ghiselin; see, e.g., Jean Rostand, *Charles Darwin*, 6th ed. (Paris, 1947). But P. Wintrebert's "thèse du vivant dans l'évolution . . . le prolongement de l'œuvre de Lamarck . . . le *Lamarckisme chimique*" can be used to illustrate certain weaknesses in Ghiselin's thesis that result from his Darwinian fetishism. See P. Wintrebert, *Le vivant, créateur de son évolution* (Paris, 1962).

the key discovery—exactly what different groups of Germans and Frenchmen were then doing.

Richet believed that there were differences between the scientific work of Germans and Frenchmen. From the viewpoint of the final results, it did not matter that Germany had the quantitative advantage over France. He illustrated by an example from his own area. An "estimable professor" at Berlin had published about 150 notes on anaphylaxis, including work done with his students, but all the facts were secondary, obscurely described, often debatable— in short, the results were nowhere proportionate to the great labor spent and the number of notes published. Similarly, in other scientific areas many German memoranda were accurate, precise, and meticulous, but often bereft of originality. The Germans frequently could not distinguish the adventitious from the essential, although the chief merit of German biologists was their meticulousness and industry.[45] But generally the classic treatises of German science were much the same as the French, for, with a few exceptions, everyone copied everyone else, simply adding accounts of the new discoveries. And if among five hundred works there was one that was excellent, the other mediocre ones could be forgiven. Finally, Richet admitted that all he had said of German works, he could say of French works as well. French works were superior because they were written in French ("prodigieusement supérieure à l'informe langue allemande"). Echoing the Fortoul argument, Richet argued that French works were written with more clarity, method, and taste. Even his statement that the role of invention was greater in French than in German science was, Richet confessed, an impression rather than a judgment, for the subject was too complex, too intricate, to permit the formulation of a solid general proposition. Science is not the patrimony of any one nation any more than of any one epoch. It would be absurd to deny the Germans their role in science because of the war. And yet they did not have either Lavoisier or Pasteur![46]

45. See also Picard, in Petit, p. 292, for the same complaint and an example from mathematics. Stanislas Meunier, "La géologie à la prussienne," ibid., pp. 264–73, attacked German geology on similar grounds, especially Suess' *Antlitz der Erde*, 2,500 pages in 3 volumes, and the work of the Sorbonne's Professor Haug, which was done in the German style and would be the *vade mecum* of students instead of Albert de Lapparent's work, done before the epidemic of Germanism hit France.

46. Richet, in Petit, pp. 348–61.

The cytologist Louis-Félix Henneguy, professor at the Collège de France, took much the same approach to German science, although he placed the Germans in a more obviously subordinate role. In the important branches of the biological sciences, Henneguy showed that one could easily line up as "fathers" or "founders" both French and Germans: Cuvier in comparative anatomy, M.-F.-X. Bichat in general anatomy, Saint-Hilaire and Lamarck in "la philosophie zoologique"; Karl Ernst von Baer in comparative embryogeny, and Theodor Schwann and M. J. Schleiden in the cell theory leading to cytology, although the work of R.-J.-H. Dutrochet and Eugène Turpin preceded that of the latter two.[47] But Henneguy hastened to emphasize that the three most eminent names in the biological sciences were Darwin, Bernard, and Pasteur, since nearly all the great discoveries in the biological sciences were developments of the doctrines of these three initiators. No German scientist had exercised a comparable influence on the evolution of the biological sciences.

In the second half of the nineteenth century, German scientists became specialists and did important work in histology, cytology, and embryology. New techniques in microscopy permitted new research like that on the comparative embryogeny of invertebrates. In spite of the work of Charles Robin, Louis-Antoine Ranvier, and others, histology had made more progress in Germany than in France. Cellular pathology as developed in Germany by Rudolf Virchow undoubtedly marked an epoch in scientific development. But the French emphasized Virchow's mistake about the dual nature of TB (an error corrected by two Frenchmen, J.-J. Grancher and L.-A. Thaon in 1872), and blamed the "disaster" of inoculation against TB on "faith in German science."[48] The French proclaimed their glory in the study of phthisis and attempted to cut Robert Koch down to a more modest role than eulogistic texts gave him: "L'histoire impartiale proclame la phtisiologie moderne, pour la plus belle part, science française: deux hommes la dominent de toute la hauteur de leur clair génie, Laënnec at Villemin, les plus grands noms qu'enregistrera la médecine mondiale au siècle de

47. Pierre Delbet (Professeur de clinique chirurgicale à la Faculté de médecine de Paris), "Sur la chirurgie allemande," ibid., p. 119. Delbet argued, similarly, that Wassermann applied to syphilis Jules Bordet's biological method of making diagnoses through studying blood serum.

48. Henri Roger, "L'évolution des sciences médicales en France et en Allemagne," ibid., p. 366.

Pasteur."[49] Henneguy noted, of course, that the way had been also prepared by others: Albrecht von Kölliker in Switzerland, Ranvier in France, and P.-J. van Beneden in Belgium. With perfected scientific equipment, great numbers of workers, and a methodical use of the results of these workers, German scientists had amassed a great number of facts, but, Henneguy insisted, had not gone beyond this patient research to achieve concomitant success in building theories based on these discoveries. The Germans, although good observers, were hindered by a lack of common sense and clarity. But even Henneguy had to admit that certain theories, like Haeckel's "gastraea" theory (an outcome of the generalization "ontogeny recapitulates phylogeny," the "fundamental biogenetic law" stressed in *Natürliche Schöpfungsgeschichte* [1867], which had gone through ten editions by 1902) and Weismann's germplasm theory of heredity (which denied the transmission of acquired characteristics) had been very successful because of their simplicity and had inspired a whole generation of researchers.

Equally astounding was the proliferation of German periodicals and the increase in the number of volumes per year. In six months, for example, the *Zeitschrift für wissenschaftliche Zoologie* published three volumes containing 2,205 pages. Most of the authors were unknown. Four of the memoranda treated the *Anodonta cellensis.* Henneguy did not explicitly condemn this but plaintively inquired about the source of the money for publication. He estimated that in anatomy, histology, cytology, embryology, and zoology Germany had at least four times more periodicals than France and that these periodicals contained ten times as much. One senses here Henneguy's envy of the fantastic energy and patience of the Germans and especially of the financial support given their activities. How useful the *Jahresberichte* were for their summary analyses of the yearly works in the different branches of the biological sciences. Even if the thankless task of reading everything was part of the German espionage system, Henneguy admitted that such a tool was sorely needed in French science. Also of value was the series of big general treatises done as collaborative works, each specialist doing the part falling within his competence. And how worthy of French emulation was the German government's financial encouragement of its scientists. But all these virtues did not offset

49. Landouzy, "Médecine allemande et médecine française," ibid., pp. 231–42.

the dangerous nature of German science, the imitation of which was ruining many young French scientists.

Henneguy's conclusion may appear to be a panicky non sequitur. But he was anxious about the spread of German methods and theories throughout the world by the large numbers of young scientists turned into disciples as a result of their German pilgrimages. Paris had been the scientific Mecca of the eighteenth and early nineteenth centuries, but now she was sharing this honor with a number of German universities, which increasingly monopolized the students coming from Europe, South America, Japan, and the United States. And even a few French scientists were publishing their researches in German scientific journals, which was the accolade needed for general acceptance of the work—a sorry snobism, in Henneguy's opinion. Worse than the bad literary style evident in the works of these young French scientists was their exaggerated confidence in German data. Acceptance of German science as infallible and sacrosanct frequently led the French scientist to reject his own contrary but correct findings. Nor could adherence to the German dogmatic models be in the best interests of science. Henneguy noted that he himself had fought against some doctrines which were by then nearly universally adopted, especially the heredity theories of Haeckel and Weismann.[50] Although Henneguy's examples were not always felicitous, his general point was solid: in spite of the cumulative nature of part of scientific knowledge, today's dogma may be tomorrow's outmoded doctrine. This healthy process should not be too inhibited by overslavish following of dogmas in one's research, even though, as Kuhn and Polanyi have argued, a substantial part of research must be done according to the dogmatic model. Henneguy thought that most of the German biologists were clever and honest observers who were often blinded

50. For some scientific criticism of Haeckel and Weismann, with heavy satire, see Yves Delage, "Histoire naturelle du DOCTUS BOCHENSIS," ibid., pp. 100–115. Félix Le Dantec, "Le bluff de la science allemande," ibid., pp. 244–50, viewed the systems of Weismann and Ehrlich as adding nothing to the wisdom of the doctus bachelierus in Molière's Malade imaginaire:

> Domandabo causam et rationem quare
> Opium facit dormire?
>
> Quia est in eo
> Virtus dormitiva
> Cujus est natura
> Sensus assoupire.

by preconceived ideas. Certain of them, like Haeckel, even doctored their data: "Il suffit, pour s'en convaincre, de regarder les planches de l'anthropogénie du célèbre Haeckel, l'inventeur du *Bathybius*, planches dans lesquelles il figure des embryons de vertébrés tels qu'ils n'ont jamais existé." Duhem also criticized Haeckel for his ruthless exploitation of the Darwinian hypothesis, demolished by "our great Henri Fabre." Duhem was especially harsh on the idea of spontaneous generation arising from nitrogenous carbon compounds. Following Arnold Brass and others, he also accused Haeckel of scientific dishonesty.[51] Many students of famous professors like Weismann were producing fantasies in order to corroborate the master's theories, a fatal result of regimenting students to perform under the master's baton. Since the Germans had revealed their intention to make science an appanage of Germany, i.e., to nationalize it, Henneguy warned that no German scientific work should be accepted unless it were subjected to a rigorous critique and verified by the French themselves. The work of all German scientists had merged with Prussian militarism.[52]

The same image of the mechanistic German style was developed by A. Chauffard, professor of clinical medicine (Paris). He frankly recognized that the best clinics in Paris were poorly equipped compared with those in Berlin and Munich. But *medical mechanization* is not all there is to clinical medicine. More important is the "esprit intérieur," a French specialty that separated the pedagogical methods of France and Germany and ultimately rested on irreducible differences separating the souls of the two races. The difference was clearly seen in the German clinical treatment of the sick as *human material*, an impious and cruel procedure that the French conscience would never accept. Adolphe Pinard expressed a similar idea: French obstetrics was the preserver of the sacred notion at the basis of humanity, "absolute respect for every human personality," which required maximum sacrifice on the part of the French in the war to ensure that "our culture annihilate *Kultur*, whose aim is to wipe out individuals and nations for its own profit."[53] In

51. Duhem, *La science allemande* (Paris, 1915), pp. 45–59. Haeckel's *Natürliche Schöpfungsgeschichte* was translated into French by the apostle of materialism, Charles Letourneau.

52. Henneguy, in Petit, pp. 206–17.

53. Le professeur Pinard (Faculté de médecine, Paris), "Puériculture—obstétricie française, obstétricie allemande," in Petit, pp. 302–24. For German criticism of German medicine in the 1920s, see Ostwald Bumke, *Eine Krisis*

Chauffard's opinion, behind the contrast of methods lay a fundamental opposition: the genius of each race is a constant and it
models in its image all the forms of national activities. The German
thought only of the material side of things and considered "organization" an end in itself. Hence Germany manufactured its
doctors as it produced its soldiers, by a mechanical process. The
French method did not place the individual under a yoke of uni-
formity but tried to develop in humanistic fashion the spark in the
student's personality that would brighten his professional life and
develop his critical faculties. Delbet went further and denounced
German surgery's "brutal technique," which was the result of the
brutality of the German character and the German cult of force.
In its totality, and especially in diagnosis, which required delicacy
of analysis and psychological skill, surgery "is essentially a French
art." As far as financial resources permitted, the German superiority
in material installations should be imitated, but that was all that
France should copy.[54]

Henri Roger, the holder of the chair of experimental and comparative pathology in the Faculty of Medicine in Paris, did not
think that scientific work was carried on in the same way in
France as in Germany. He put forward two systems or models of
scientific research for the two countries, covering at least the medical sciences. German medical researchers were applying to the
study of human sickness simple methods confined to the laboratory.
Here was where Germany thought she had triumphed: method and
organization in the laboratories, subsidies for scientists, expensive
installations, and facilities open to scientists of all countries. In the
laboratory ruled by the professor each worker followed the discipline model of the army in pursuing a line of research dictated by
the professor. The disadvantage of this despotic system was that
it restricted the mind; the advantage was that research could be
carried out without any great intellectual prowess. Germany had
the very real merit of knowing how to use the most mediocre minds,
substituting patience, care, and perseverance for brilliance. The
same point was made by Delbet: "Ce qu'on appelle la méthode

der Medizin: Rede gehalten bei der Übernahme des Rektorats am 24. November 1928 (Munich, 1929), and Ringer, The Decline of the German Mandarins,
pp. 385–86.
    54. A. Chauffard, "L'enseignement de la clinique médicale en France et en
Allemagne," in Petit, pp. 56–68, and Delbet, ibid., pp. 117–37.

allemande permet d'utiliser les cerveaux les plus médiocres, de produire de gros travaux sans idée." Delbet argued that what was most extraordinary about the fame of German surgery was that it had been done mostly by French surgeons. His argument went further: surgery had been transformed by anesthesia, hemostatic procedures, and antisepsis, areas in which the Germans had made no great discoveries but which were the preserves of the Americans, English, and French.[55] Roger admitted, of course, that there were great scientists in Germany, but genius was as rare in Germany as in France.

The French model of scientific research was quite different. Laboratories, at least in the Faculty of Medicine, were less well subsidized and workers fewer. Most young people turned away from scientific research because of the extensive preparation for competitive examinations. Those who found the time to devote themselves to personal studies were too independent in mind and too rambling in disposition to spend years in a narrow specialty in order to study a single detail. The diversity of their interests unfortunately diffused their efforts but at the same time broadened their knowledge. Nearly all handled clinical medicine, histology, and experimentation. Although their work frequently lacked precision, it did possess originality. Roger subscribed to the typical French view that subordinate laboratory workers in France were free and independent, whereas they were under the master's baton in Germany. This is not entirely a French fantasy, perhaps, for the American scientist Eben Norton Horsford thought that his reception by Liebig was "rather that of a military officer than of a scientific man." Roger did not think it possible to say that one system produced better results than the other, for no one had the criteria to judge the value of scientific works nor to discover their long-range results.[56] German discipline did not suit the French mind and French independence did not suit the German mind.

After an enumeration of French glories in medicine, Roger noted that the success of German science was partially due to its efficient propaganda. Dastre also saw the German book and periodical industry as a powerful propaganda instrument, especially the Zeit-

55. Henri Roger, ibid., pp. 363–74; Delbet, ibid., pp. 118–35.
56. See Michael Polanyi, *Personal Knowledge* (Chicago, 1958, 1962; New York, 1964), pp. 134–42, for the argument that "Though not definable in precise terms, scientific value can as a rule be reliably assessed."

schrift für physikalische Chemie, which published fine non-German works that then redounded to the scientific credit of Germany.[57] France modestly and mistakenly did not follow the German custom of putting its discoveries in the limelight. Content to accumulate new facts, spread general ideas, develop big synthetic conceptions, France did not declare to the world that she had monopolized science.[58]

57. Dastre, in Petit, p. 91.
58. Roger, ibid., pp. 364–74.

3. Pierre Duhem as Propagandist: A Subtle Revision

THE MOST detailed, penetrating comments on German science came from the physicist and historian and philosopher of science Pierre Duhem, who probably had a more solid grasp of the technical part of German science, at least in physics, chemistry, and mathematics, and of its historical development than any other commentator.[1] Judging science under the rubric of a national mentality was not new for Duhem. In 1893 he had written a penetrating article on English physics and theory, expanded in chapter 4 of his famous work *La théorie physique—son objet—sa structure*. This book was translated into German by Friedrich Adler and given an enthusiastic endorsement by Ernst Mach: "Möge das Buch nach Verdienst auch in Deutschland Anerkennung finden, aufklärend und fördernd wirken!"[2] Since the epistemological views enunciated by Duhem in that chapter formed the basis of his later comments on the national characteristics of science, it is useful to have some idea of Duhem's quasi-Pascalian views on types of mentality, especially as applied to national groups.

The structure of Duhem's argument rests on his view that a physical theory is produced through the processes of abstraction and generalization. First, the mind analyzes a large number of

1. See Pierre Duhem, *Revue des deux mondes* 25 (February 1, 1915): 657–86; *La science allemande*; and "Science allemande et vertus allemandes," in Petit, pp. 138–52. The article in the *Revue des deux mondes* was reprinted as a supplement to the four lectures published in *La science allemande*.
2. The article of 1893 appeared in the *Revue des questions scientifiques*. See chapter 4, "Les théories abstraites et les modèles mécaniques," of *La théorie physique* (Paris, 1906, 1914, 1933); translated into English by Philip P. Wiener, *The Aim and Structure of Physical Theory* (Princeton, 1954; paperback edition by Atheneum, 1962), and into German, *Ziel und Struktur der physikalischen Theorien* (autorisierte Übersetzung von Dr. Friedrich Adler, Privatdozenten an der Universität Zürich). Mit einem Vorwort von Ernst Mach (Leipzig, 1908). On Mach see Robert S. Cohen and Raymond J. Seeger, eds., *Ernst Mach: Physicist and Philosopher* (Reidel, 1970).

facts and summarizes their common and essential elements in a
law ("a general proposition tying together abstract notions"). Second, after pondering a group of laws, the mind substitutes for
the group a small number of very general judgments containing
some very abstract ideas. The mind chooses these *primary qualities*
and formulates these *basic hypotheses* in such a way that a long
but unerring deduction can draw from them all the laws belonging
to the group studied: "This system of hypotheses and deducible
consequences, a work of abstraction, generalization, and deduction,
constitutes a physical theory. . . ."[3] The dualistic operation of abstraction and generalization effects a double economy of thought
in substituting, first, a law for a multitude of facts and, second, a
small group of hypotheses for a large set of laws. Reducing facts to
laws and laws to theories is an economy of thought that makes it
much easier for abstract minds to learn physics. But Duhem recognized that not all powerful minds are abstract minds. Some minds
have a wonderful talent for producing in their imagination a clear
picture of a great many different objects, if they can be perceived
by the senses. These *imaginative minds* (*esprits imaginatifs*) find
abstraction and deduction difficult and, since they regard abstract
theory not as an intellectual economy but as a painful procedure of
dubious utility, they build their physics on a different model. Only
abstract minds will readily accept the Duhemian conception of
physical theory as the true form in which to represent nature.
Duhem summed up his case for mental types by quoting Pascal:
"Thus there are two kinds of mind: one goes rapidly and deeply
into the conclusions from principles, and this is the accurate mind
(*l'esprit de justesse*). The other can grasp a large number of principles and keep them distinct, and this is the mathematical mind
(*l'esprit de géométrie*). The first is a powerful and precise mind,
the second shows breadth of mind. Now it is quite possible to have
one without the other, for a mind can be powerful and narrow, as
well as broad and weak."[4] Within the Pascalian definitions, Du-

3. Duhem-Wiener, *Physical Theory*, p. 55. Cf. the "typical presentation of
the so-called hypothetico-deductive, or inductive, method of science" given by
Friedrich Dessauer in *Eranos-Jahrbuch* 14 (Zürich, 1946), summarized in
Gerald Holton, ed., *Science and Culture. A Study of Cohesive and Disjunctive
Forces* (Boston, 1965), pp. 282 ff.
4. Pascal, *Pensées*, trans. A. J. Krailsheimer (Baltimore, 1968), pp. 209–10.
Two pages on the mathematical and intuitive minds follow. Wiener translates
esprits imaginatifs as visualizing minds. See also translator's note on p. 60 of
Duhem-Wiener, *Physical Theory*.

hemian abstract physical theory would win strong but narrow minds and be rejected by broad but weak minds.[5]

Although the ample type of mind can be found in all countries, Duhem argued that it is endemic in England: "This opposition between the French mind, strong enough to be unafraid of abstraction and generalization but too narrow to imagine anything complex before it is classified in a perfect order, and the ample but weak mind of the English will come home to us constantly while we compare the written monuments raised by these two people."[6] Heroes of Shakespeare and Corneille, or of the novels of Dickens and Eliot compared to those of Loti, provided convenient examples of the two types of mentality prevalent in France and England. In philosophy Descartes' *Discours de la méthode* and Bacon's *Novum Organum* clearly showed the strong and restricted mind of the geometer as compared with the chaotic, imaginative faculty of Bacon, "with its taste for the concrete and practical, its ignorance and dislike of abstraction and deduction," all of which gave the Baconian philosophy an industrial aim.[7] The opposition between the French and English mentalities could also be seen in their respective legal systems and social structures. But Duhem was chiefly concerned with physics.

As is clear from the discussion on physical chemistry by the Sorbonne faculty in 1895, the concept of national styles of scientific thinking was prevalent in the French scientific community in the period under consideration. But it was not limited to the French. In the inaugural address to the mathematics and physics section of the congress of the British Association for the Advancement of Sciences in 1892, the English physicist Arthur Schuster used the same concept. Schuster thought that each nation had its speciality. No country could compete with France in precise measurement, a field then associated with the names of H.-V. Regnault and Emile Amagat; it made sense to situate the Bureau international des poids et mesures in Paris. German scientific genius lay in an entirely different area. Schuster was skeptical about the possibility of copying the Germans: "L'allemand excelle à pousser une théorie jusqu'à ses conclusions logiques et à les soumettre à l'épreuve expérimentale;

5. Duhem, *La théorie physique*, pp. 77–81; Duhem-Wiener, *Physical Theory*, pp. 55–57.
6. Duhem-Wiener, *Physical Theory*, p. 64.
7. Ibid., pp. 66–67.

je ne crois pas que les efforts que l'on pourra faire pour transplanter dans notre pays les travaux de recherches des universités allemandes puissent être couronnés de succès. Ne vaut-il mieux laisser chaque pays s'attribuer la part du travail qui convient le mieux à son caractère et à son éducation? Est-il sage de porter remède à quelques points faibles, de boucher quelques trous, si la terre pour remplir ces trous doit être enlevée aux collines qui émergent au-dessus du niveau général?" Schuster thought that England had excelled in mathematical physics, although she had done nearly as well in astronomy, chemistry, and biology. What especially distinguished England from other countries was the predominant role played by scientific amateurs, a group which the modern system of education tended to eliminate.[8]

One of the clearest scientific statements on the distinctive characteristics of the French mind and of French science came from the geologist Albert de Lapparent.

L'esprit français a soif de clarté. Il veut que les choses lui soient presentées nettement, même quand la netteté de l'exposition devrait dépasser un peu ce que semblent autoriser les notions réellement acquises. D'autre part, il ne peut se contenter de connaître le *comment* des phénomènes. C'est un besoin pour lui d'en apercevoir le *pourquoi*, c'est-à-dire de les rattacher les uns aux autres, par ces relations de cause à effet dont l'enchaînement logique constitue ce qu'on appelle des théories. [There is no point, Lapparent continued, in discouraging this penchant of the French mind by pointing out that a doctrinal edifice has only a limited duration, that observation reveals a phenomenon theory cannot accommodate. The French mind knows that this development is just an evolution, not an overthrow, and that a modification of the theory will accommodate new facts.] . . . Un instinct sûr avertit l'esprit français que la vraie science a pour objet principal non la connaissance des résultats d'expérience, mais l'intelligence des rapports qui les unissent. Tandis que l'observation perfectionne ses méthodes, et introduit une précision croissante dans l'expression des faits constatés, le savant se sert de ces progrès pour mieux définir les rapports déjà entrevus, de sorte que peu à peu les lignes maîtresses de l'édifice doctrinal se dégagent avec une netteté grandissante.

8. Arthur Schuster, "Histoire des sciences. Les auxiliaires de la science: Amateurs et théoriciens," *Revue scientifique* 50, no. 16 (1892): 481–87.

C'est à ce point de vue qu'on a vraiment le droit de dire qu'il existe une science française; car si la connaissance de phénomènes est une de sa nature, et n'a pas à compter avec les distinctions de race ou de nationalité, l'idée qu'on fait des choses n'est nullement indifférente au progrès de l'observation elle-même, qu'elle guide en l'orientant vers des voies fécondes. Or tandis que, dans d'autres pays, on se contente volontiers de recueillir des faits, évitant avec une défiance systématique toute tentative de les réunir en théorie, chez nous on professe de longue date ce qu'exprimait si bien M. H. Poincaré dans son discours au congrès de physique de 1900. . . . "Le savant doit ordonner: on fait la science avec les faits comme une maison avec des pierres; mais une accumulation de faits n'est plus une science qu'un tas de pierres n'est une maison."[9]

The French were always astonished to find the English physicists using models in the exposition of a theory. For the French or German physicist the theory of electrostatics, for example, "constitutes a group of abstract ideas and general propositions, formulated in the clear and precise language of geometry and algebra, and connected with one another by the rules of strict logic. This wholly satisfies the reason of a French physicist and his taste for clarity, simplicity, and order."[10] But this mode of representation, declared the English physicist Lodge, could not enable him to "form a mental representation of the phenomena which are really happening." A mechanical model was needed. But Duhem disdained the materialization of Lodge's mechanical model: "We thought we were entering the tranquil and neatly ordered abode of reason, but we find ourselves in a factory." Thomson had also declared that "I never satisfy myself until I can make a mechanical model."[11] Committed to the mechanical explanation of physical phenomena, the English found the use of models so vital in physics that they confounded the model with the understanding of the theory itself. Duhem saw the French and continental scientists as the followers of the "purely abstract theory highly regarded by Newton," although Gassendi had an am-

9. A. de Lapparent, "L'évolution des doctrines cristallographiques," *Revue de l'Institut catholique de Paris*, January–February 1901, no. 1, pp. 1–24, and March–April 1901, no. 2, pp. 123–43. At all times the merit of French scientists was to have been guided by such a principle. Lapparent thought this especially true in crystallography, "by its origins, the most French."
10. Duhem-Wiener, *Physical Theory*, p. 70.
11. Ibid., pp. 70–71.

ple but shallow mind, as came out in his skirmish with Descartes.[12] Certain aspects of the English "predilection for explanatory and mechanical theories" were present on the continent, but although the continental sense of abstraction lapsed, as in the seventeenth-century reduction of matter to geometry, a triumph of imagination over reason, it never fell asleep completely, and reason always recovered its rights.

In his discussion of mathematical and intuitive minds Pascal had noted that it was rare for mathematicians to be intuitive or the intuitive to be mathematicians. Following Pascal, Duhem viewed ampleness of mind as "the characteristic quality of the genius of the pure algebraist," whose skill, not concerned with analyzing abstract ideas and treating the scope of general principles, is an "ornament of the imaginative faculty."[13] Among English mathematicians this skill was prevalent and manifested itself in "the Englishman's predilection for the diverse forms of symbolic calculation." Thus the English scientist in his works made use of the "complex and shorthand languages" of symbolic algebras, the calculus of quaternions, and vector analysis, whereas "French and German mathematicians do not learn these languages readily; they never succeed in speaking them fluently or, above all, in thinking directly in the forms which constitute these languages."[14] For the French or German scientist "a physical theory is essentially a logical system." In developing a theory the French or German physicist used the algebraic language of a theory to replace the syllogisms used to develop it. But Maxwell's exposition of his electromagnetic theory followed the English pattern of ignoring definition and set up equations for a physical theory. Making no comparison between his algebraic model and the physical laws it had to imitate, he followed the model and combined the electrodynamic equations. In Hertz' words, "Maxwell's theory is the system of Maxwell's equations." Einstein later noted what probably bothered Duhem: since Maxwell's equations could not be interpreted mechanically, it was necessary to abandon the premise that mechanics was the foundation of all physics. The model satisfies the imagination, not reason, for there is no "series of deductions, in the algebraic transformations, from clearly formulated hypotheses to empirically verifiable

12. Ibid., p. 87.
13. Ibid., p. 76.
14. Ibid., pp. 76–77.

conclusions."[15] Unless the French or German physicist was aware of this, he would be puzzled as to what Maxwell's theory was, a question that in its continental meaning had no answer for Maxwell's typically English theory: "A gallery of paintings is not a chain of syllogisms."

This last remark will probably intrigue the contemporary reader, who may wonder why Duhem merely moved on the periphery of a question tantalizing some minds today: what is the connection between creativity in art and in science? It is a pity that Duhem, who had considerable talent for drawing, did not answer the question with his customary clarity, for some now see a close relation of the issue to the new direction science was then taking. Had he done so, it is possible, considering his views on the nature of science, that his answer would not have differed in its essentials from that of a contemporary historian of science, Thomas S. Kuhn, who does not think that art can be distinguished from science by any of the classic dichotomies like value–fact, subjectivity–objectivity, and intuition–induction.[16] But Kuhn argues that although the scientist and the artist are "guided by aesthetic considerations and governed by established modes of perception . . . the artist's goal is the pro-

---

15. Ibid., pp. 79–80. Translation slightly modified. See also pp. 80–86 for further comment, including substantiation of the point by quoting of Poincaré's similar reaction to Maxwell's *Treatise on Electricity and Magnetism.* Maxwell's "lack of concern for all logic and even for any mathematical exactitude" is dealt with in Duhem's *Les théories électriques de J. Clerk Maxwell: Etude historique et critique* (Paris, 1902). See Einstein's autobiographical notes in Paul A. Schilpp, *Albert Einstein: Philosopher-Scientist* (Evanston, Ill., 1949).

16. For an extended discussion of these issues, see the articles of Ackerman and Hafner and the commentary of Kuhn in *Comparative Studies in Society and History* 2, no. 4 (October 1969): 373–412. See also Claude Ambroise, "Marinetti, le chantre du monde mécanique," *Le Monde,* May 9, 1970; Noël Mouloud, "Les formes dans la science et le formalisme," *La Pensée,* no. 150 (April 1970); and Gyorgy Kepes, "The Visual Arts and the Sciences: A Proposal for Collaboration," in Holton, ed., *Science and Culture,* pp. 145–62. Cf. Holton, "The Thematic Imagination in Science," in *Science and Culture,* pp. 88–108; he brings science and humanistic scholarship together in the area of "the formation, testing, and acceptance and rejection of hypotheses." Charles Edward Gauss, *The Aesthetic Theories of French Artists, 1855 to the Present* (Baltimore, 1949), presents some interesting and very general parallels between artistic and scientific developments and points out how epistemological questions influenced art as well as science. It is curious that no one refers to the fact that quite a few scientists, like Duhem, Ostwald, and Jules Jamin, had considerable artistic talent. Jamin's son, like Victor Regnault's, was a well-known artist. In Duhem-Wiener, *Physical Theory,* pp. 24, 30, 288, there are a few references to the aesthetic aspect of science.

duction of aesthetic objects . . . whereas for the scientist the aesthetic is a tool for solving a technical puzzle." Whether or not one accepts this distinction between the activities of the artist and the scientist, it can be argued that it is quite misleading in any consideration of the "picture of nature" developed in theoretical physics since Maxwell. Physicists like Schrödinger "worked from a more mathematical point of view" in quantum mechanics than did Heisenberg, who, "keeping close to the experimental evidence about spectra . . . found out how the experimental information could be fitted into . . . matrix mechanics."[17] According to Dirac, Schrödinger, in extending de Broglie's idea of a wave-particle, got "a very beautiful equation . . . for describing atomic processes . . . by pure thought, looking for some beautiful generalization of de Broglie's ideas. . . ."

This is the type of thing Duhem criticized in "the English manner of dealing with physics," that, in Dirac's words, "it is more important to have beauty in one's equations than to have them fit [the] experiment," since discrepancies are frequently "cleared up with future developments of the theory." But experiments can also be beautiful, as Einstein saw: "I always think of Michelson as the artist in science. His greatest joy seems to come from the beauty of the experiment itself, and the elegance of the method employed."[18] Since "fundamental physical laws are described in terms of a mathematical theory of great beauty," Dirac thinks that "the next advance in physics will come about along these lines: people first discovering the equations and then needing a few years of development in order to find the physical ideas behind the equations." But Duhem, like Poincaré, thought it dangerous to separate these two complementary components of science, even for a temporary pe-

17. P. A. M. Dirac, "The Evolution of the Physicist's Picture of Nature," *Scientific American* 208, no. 5 (May 1963): 45–53. Since Dirac leaves a large role for experimental physicists and Kuhn admits that "the development of mathematics resembles that of art more closely than that of science," it may be argued that, on some grounds, they are not as far apart as it seems on first analysis. But the emphasis of each is clearly different. Kuhn's distinction between paradigm and theory and his parallels between paradigm and picture, rather than theory and style, must also be considered. Support for Kuhn's argument regarding art comes from Brandt Kingsley: "Nous ne sommes plus les mêmes et ces maîtres [du Bauhaus] nous ont appris, à nous, les professeurs d'art, que ce qui compte, ce n'est pas le résultat final, mais la recherche, et pas le produit mais le procédé." *Le Monde*, June 11, 1970.

18. Gerald Holton, "Einstein, Michelson, and the 'Crucial Experiment,'" *Isis* 60, part 2, no. 202 (1969): 156.

riod.[19] Poincaré believed strongly in the need for experiment: "Rien n'est plus dangereux, en principe, que d'admettre des thèses ès-sciences physiques ne contenant pas d'expériences nouvelles."[20] Here Duhem sensed himself out of sympathy with some contemporary trends in physics. This is superbly illustrated by Jacques Hadamard: "having . . . obtained a simple result [the 'composition theorem'] which seemed to me an elegant one, I communicated it to my friend . . . Duhem. He asked me to what it applied. When I answered that so far I had not thought of that, Duhem, who was a remarkable artist . . . compared me to a painter who would begin by painting a landscape without leaving his studio and only then start on a walk to find in nature some landscape suiting his picture. This argument seemed to be correct, but . . . I was right in not worrying about applications: they did come afterwards."[21] The basis of Hadamard's confidence was esthetic: "Je m'étais fié sans réserve au sentiment de beauté que me donnait mon énoncé, et ce sentiment ne m'avait pas trompé."[22]

Duhem recognized that in the late nineteenth century "the English manner of dealing with physics has spread everywhere with extreme rapidity. Today it is customarily used in France as well as in Germany."[23] This reduction of physical theory to a col-

19. The case of Poincaré is complex. For development of the idea that part of the Poincarist philosophy was that "La valeur de la science tient uniquement à l'esthétique qu'elle enferme," see J. Segond, *Art et science dans la philosophie française contemporaine* (Paris, 1936), pp. 55–67.

20. "Rapport sur la thèse de M. Charles Nordmann" (Poincaré, rapporteur), June 13, 1903, AN, F17 13248. See also Jules Tannery, "Mathématiques pures," in *De la méthode dans les sciences* (Première série, Paris, 1920), p. 63, who, after referring to "la conception mystique" in Hermite, declares, "Je suis, toutefois, bien convaincu qu'aux spéculations les plus abstraites de l'analyse correspondent des réalités qui existent en dehors de nous et parviendront quelque jour à notre connaissance."

21. Jacques Hadamard, *An Essay on the Psychology of Invention in the Mathematical Field* (Princeton, 1949), p. 128. On p. 127, he relates that during the defense of his doctoral thesis Hermite "observed that it would be most useful to find applications," which Hadamard did not then have. See Duhem-Wiener, *Physical Theory*, pp. 139–41. For an appreciation of Duhem from the viewpoint of contemporary science, see Louis de Broglie's introduction to Duhem-Wiener, *Physical Theory*, and Donald G. Miller, "Pierre Duhem, un oublié," *Revue des questions scientifiques*, 138 (October 1967): 445–70; this article appeared originally in English in *Physics Today* (December 1966).

22. R. Taton, *Causalités et accidents de la découverte scientifique. Illustrations de quelques étapes caractéristiques de l'évolution des sciences* (Paris, 1955), p. 23. See pp. 22–23 for a variant of this story.

23. Duhem-Wiener, *Physical Theory*, p. 87. It can be argued that "the

lection of models resulted chiefly from the imitation of Maxwell by continental disciples. "Among the best of those who have helped promote such a fashion of treating mathematical physics" was Hertz, whose theory of electrodynamics was built on Maxwell's equations.[24] Helmholtz' electrodynamic theory, on the other hand, was fashioned on the old continental model: it proceeded logically from the solid principles of electrical science, formulated in equations not tainted by the paradoxes often found in Maxwell's. It explained all the facts accounted for by the equations of Maxwell and Hertz and did not clash with reality. Reason, urged Duhem, prefers Helmholtz but "imagination prefers to play with the elegant algebraic model" of Hertz, and of Heaviside and Cohn.[25] Among the various reasons for the popularity of the English scientific model, Duhem curiously assigned primary importance to the industrial demands, connected with English physics, rather than to any desire to find beautiful equations. Neither the industrialist nor the engineer could afford the time for abstract physics. Science was confused with industry, and the dusty, smoky, and smelly automobile was viewed as the triumphal chariot of the human mind. In a section entitled "Is the Use of Mechanical Models Fruitful for Discoveries?" Duhem argued that "The share of booty it has poured into the bulk of our knowledge seems quite meager when we compare it with the opulent conquests of abstract theories." The most invidious consequence of imitating the English was that it was easiest to ape their defects; so the snob who imitated the English would probably end up with the worst mental qualities of the English and the French, that is, with a weak and narrow mind—a false mind.[26] Each type of mind has its virtues and vices: "The best means of promoting the development of science is to permit each

---

absence of intelligible models in quantum physics . . . [makes] something like Duhem's position . . . the accepted philosophy underlying modern physical theory." For a partial defense of the idea "that without models theories cannot fulfil all the functions traditionally required of them, and in particular that they cannot be genuinely predictive," see Mary B. Hesse, *Models and Analogies in Science* (Notre Dame, Indiana, 1966), especially the debate between two modern disciples of Duhem and the English physicist N. R. Campbell. But see D. H. Mellor, "Models and Analogies in Science: Duhem *versus* Campbell," *Isis* 59, no. 198 (1968): 282–90, for the argument that Duhem and Campbell are in essential agreement on the issue.

24. Duhem-Wiener, *Physical Theory*, p. 90.
25. Ibid., pp. 90–91.
26. Ibid., pp. 92–99.

form of intellect to develop itself by following its own laws and realizing fully its type. . . . In a word, do not compel the English to think in the French manner, or the French in the English style." This principle of *intellectual liberalism* was advocated by Helmholtz. The mad desire to copy English physics transformed the very harmonious and very logical theoretical physics constructed by the French into a disgraceful and confused heap of illogicalities and nonsense. Duhem argued that French intelligence could be safeguarded against the threats of both English and French science by studying the classics of science: "Prenez-y bien garde. Admirez le génie anglais; ne l'imitez point. Pour procéder à l'anglaise dans la recherche de la vérité, il faut avoir l'esprit fait à l'anglaise; il faut posséder cette extraordinaire faculté d'imaginer simultanément une foule de choses concrètes sans éprouver le besoin de les ranger, de les classer; or il est très rare que le Français soit doué de cette faculté." Science had changed but not the way of doing it well. Duhem's pedagogical recommendation set forth a policy carefully avoided in contemporary scientific education.[27]

In his long wartime article on German science in the *Revue des deux mondes*,[28] Duhem warned that the author of such an essay must guard against hard and fast conclusions because essentially and in its perfect form science must be absolutely impersonal. If no discovery can carry the signature of its discoverer, it is not possible to declare in what country the discovery occurred. But this perfect form can only be attained in ideal circumstances—an exact following of the various methods of discovery. The multiple faculties of reason have to play their role without usurping the role of any other faculty. Since these faculties do not exist in perfect equilibrium in any individual, the science produced by individuals does not possess the harmonious aspects of an ideal science. Thus it is possible, argued Duhem, to indicate which people have produced various doctrines. As is clear from his treatment of English and French science, he belived that each nation had its own peculiar mental characteristics and therefore produced a science with distinguishing traits. Duhem hastened to add two qualifications to the

27. Ibid., pp. 92–95. Also *La science allemande*, pp. 91–92. For a model of contemporary scientific education, see Thomas S. Kuhn, "The Essential Tension: Tradition and Innovation in Scientific Research," in *The Third (1959) University of Utah Research Conference on the Identification of Creative Scientific Talent* (Utah, 1959), pp. 164–65.
28. See note 1 above.

idea of assigning national characteristics to a people. First, judgments on the intellectual characteristics of a people can never be verified because they are not universal. Not all Englishman are English types; theories conceived by the English did not have all the characteristics of English science, and France had intellectuals who thought in the English manner. Second, the national character of an author could be seen only in those characteristics that caused his doctrines to deviate from the idea of a perfect type. It is in its defects, then, that Science becomes the science of any people or national group. The marks of genius of any nation are especially evident in second-rate scientific works produced by mediocre thinkers. Very often the great masters possess a reason in which all the faculties are so harmoniously blended that their finished doctrines are exempt from individual idiosyncrasy and from all national character. No trace of the English mind could be found in Newton and no trace of the German could be found in Gauss or Helmholtz; in such works only the genius of humanity is evident.

In all science that has put on mathematical garb, Duhem detected two types of procedure: the tactic that obtains principles and the tactic that arrives at conclusions. The method of going from principles to conclusions is the deductive method followed with rigorous accuracy. The method leading to the formulation of principles is more complex and more difficult to define, as can be seen by examining the history of axioms in geometry from Euclid to Hilbert. Even more complex is the choice of hypotheses for the foundation of a doctrine in experimental science or of a physical theory.[29] The choice of a theory in mechanics or physics could not be accomplished by the deductive method, which is only an inflexible and insufficiently penetrating help. Even more than the mathematician, the physicist, in choosing his axioms, needs a faculty distinct from the *esprit géométrique*; he needs the *esprit de finesse*. The geometrical or mathematical mind and the subtle mind do not advance at the same pace. The progress of the geometrical mind has inflexible rules imposed on it which are not of its own creation; each of the propositions it unfolds has its position in the pattern determined in advance by a necessary law. In its operations this type of mind resembles an army on parade. Whereas the *esprit géométrique* derives the power of its deductions from the accuracy of its advance, the penetration of the subtle mind (*esprit de finesse*) derives en-

29. *Revue des deux mondes*, pp. 659–60.

tirely from the spontaneous flexibility with which it moves. No immutable precept determines its untrammeléd advances; although it operates within the framework of order, it is an order prescribed for itself and modified according to circumstances. The progress of the subtle mind resembles a group of sharpshooters assaulting a difficult position under a general order to capture the position, although each is left free as to the action he should take to interpret the order and achieve the prescribed aim.[30]

It was on the basis of this comparison between the *esprit de finesse* and the *esprit géométrique* that Duhem established the characteristic of German science that distinguished it from French science. Most French science is characterized by use of the *esprit de finesse*, which, made impatient by the restrictions and slowness of the *esprit géométrique*, becomes aggressive to the extent of sometimes encroaching upon the latter's prerogatives. German science is characterized by the *esprit géométrique*, which sometimes ventures into areas not within its proper ken. Duhem thought that the geometric mind might better be called the algebraic mind. To construct algebra, which rests on a small number of simple propositions, one needs the ability to follow unerringly long, complex series of detailed rules of logic. One also needs the ability to perceive in a complex algebraic expression the various transformations that can be made of it according to the rules of calculation and thus arrive at the formulae one wants to discover. This is not a power of reasoning but a talent for combination. Although there were, Duhem admitted, some German mathematicians who possessed to a high degree the ability of combination, it was not in this area that they excelled. It was easier to find this type, a Hermite, a Cayley, or a Sylvester, in England or France.[31] This distinction was a most fortunate and necessary one if Duhem were to distinguish German science from English science, the structural similarity of which would be a most unhappy argument to make in 1915. In his commentary in 1893 on English science, especially physics, Duhem had emphasized the algebraic skill of English scientists in contrast to

30. Ibid., pp. 660–61.
31. Henry E. Guerlac, in Earle, p. 82, comments that "Very possibly France's greatest contribution has not been in natural science at all, but in mathematics. It could be argued that in no other scientific field can France produce an aggregate of names as impressive as those of Viète, Descartes, Pascal, Lagrange, Laplace, Cauchy, Hermite, Henry Poincaré, Lebesgue, and d'Ocagne." Guerlac admits that this may be an exaggeration.

French and German scientists. He had set up at that time an opposition between English science and continental science, usually referring to French and German scientists as if their basic approaches and fundamental assumptions did not differ, although there were some significant German exceptions like Hertz.

The distinction of German algebra, according to Duhem, lay in its power to produce and follow with the greatest accuracy long and complex chains of reasoning, such as were evident in the work of Weierstrass and Cantor. Because of his attention to minutiae Schwartz could boast that he was the only mathematician who had never made a mistake. One of Duhem's friends, who had followed Schwartz' courses at Göttingen, complained to Duhem about the agony a Frenchman underwent in following such painstaking and trifling mathematical sorites.[32] This submission of their *esprit géométrique* to the rules of deductive logic had enabled the German mathematicians to make very useful contributions to the perfection of analysis and to rid algebra of paralogisms sometimes introduced by the swift and summary intuition of mathematicians in whom the *esprit de finesse* predominated. The disadvantages of the *esprit géométrique*, however, obviously outweighed its advantages: it engulfed science with idle and tedious discussions about trifles, and it stifled the spirit of discovery, dangers of which German mathematicians like Felix Klein were aware. The intuition that discovery precedes demonstration in mathematics is an attribute of the *esprit de finesse*. Some German mathematicians saw that its stifling by an overgrowth of the *esprit géométrique* could have dire consequences for original discovery.[33] And the German mind was especially susceptible to the intellectual imperialism of the *esprit géométrique*. Since algebra subjected reason to the iron discipline of syllogistic laws and rules of calculation, no science was more suited to the German mind, proud of its geometric accuracy but lacking in subtlety. So the Germans tried to bring all science to a form as close as possible to that of algebra, which meant, for example, the reduction of geometry to a branch of analysis. The doctrine of Riemann, as seen in the work of genius of his geometric mind, *Über die Hypothesen welche der Geometrie zu Grunde liegen*, was a rigorous algebra, for all its theorems were rigorously deducted from its stated postulates, thus satisfying the geometric mind: "It is

32. Duhem, *La science allemande*, pp. 9–10.
33. Duhem, *Revue des deux mondes*, pp. 662–63.

not a *true geometry* because in laying down its postulates it is not concerned with whether their corollaries agree totally with judgment drawn from the experience that makes up our intuitive knowledge of space; it also shocks common sense."[34] This justly famous work of German science was, Duhem concluded, a remarkable example of the procedure by which the geometric mind of the Germans transformed every doctrine into a sort of algebra.

Duhem believed "that physical science flows from two sources: the certainty of common sense and the clarity of mathematical deduction."[35] He was also keenly aware of the importance of the historical method in physics: "To give the history of a physical principle is at the same time to make a logical analysis of it."[36] The hypotheses upon which a theory in mechanics or mathematical physics rests result from the interaction of a great many factors, not all of which are scientific.[37] Only a very subtle mind could unravel all the twisted threads that had been woven into the development of these hypotheses. But in the mathematical physics of Gustav Kirchhoff no trace of the historical elements could be found: each hypothesis was presented *ex abrupto* in its abstract and very general form; a physical theory was only a sequence of algebraic deductions. Hertz' mechanics was developed under the austere aegis of the geometric mind: there is no mention of mechanics from Buridan to Gauss, but simply systematic abstraction. Often such a procedure had happy consequences, preferable to a frequent French illusion that a physical principle was demonstrated when it was made attractive. The pure algebraism of German theories was marvelously suited for producing what Mach called an economy of thought by reducing a host of experimental laws to a small number of theoretical postulates. Duhem concluded that both the geometrical mind and the subtle mind were vital to science: a physics vitiated by an excess of the one was cured by an excess of the other. Belladonna

---

34. Ibid., p. 668. See pp. 664–65 for comments on how the works of Gauss, Bolyai, and Lobachevski on whether Euclid's axioms are truly independent of one another fit into the activity of the *esprit géométrique*.

35. Duhem-Wiener, *Physical Theory*, p. 267.

36. Ibid., p. 269.

37. See, for example, Thomas S. Kuhn, *The Copernican Revolution. Planetary Astronomy in the Development of Western Thought* (Cambridge, Mass., 1957), and C. C. Gillispie, "Intellectual Factors in the Background of Analysis by Probabilities," in A. C. Crombie, ed., *Scientific Change* (London, 1963), pp. 431–53.

and digitalis neutralize each other's effects, but they are both poisonous plants![38]

Like many other French commentators, Duhem complained that German scientists did not regard very highly the demands of common sense. Rather they seemed to take a malicious pleasure in scandalizing the *esprit de finesse* by building their meticulously arranged system or apparatus upon the foundation of an affirmation contradicting the surest principles of logic. Drawing upon his vast knowledge of the history of science, Duhem traced the development of this scientific *Schadenfreude* back to Nicolas of Cusa's *De docta ignorantia*. Duhem also attributed the success of Hegelianism in Germany to the pleasure that the geometric mind found in the purely deductive method, without having its common sense shocked by the idea of an identity of opposites. Indeed, the absence of a mutual interpenetration between science and life was glaringly evident in the idealist philosophy. With his genius for illustration and caricature, Duhem sketched the idealist philosopher: "Dans sa chaire d'Université, il dénie tout réalité au monde extérieur, parce que son esprit géométrique n'a pas rencontré cette réalité au bout d'un syllogisme concluant. Une heure après, à la brasserie, il trouve une satisfaction pleinement assurée dans ces pesantes réalités que sont sa choucroute, sa bière et sa pipe."[39] The absence of the *esprit de finesse* in German science permitted a gulf to develop between the development of ideas and the observation of facts: ideas pullulated from one another in deductive fashion, proud of contradicting common sense; and common sense was left to handle reality and ascertain the facts without reference to theory.[40]

The incoherent duality of German science was evident in one of

38. Duhem, *Revue des deux mondes*, pp. 669–71.

39. Ibid., p. 673. Duhem gave a long quote from Boutroux' article in the *Revue des deux mondes* on the German separation of the scientist and the man.

40. Ibid., p. 674; but cf. Duhem-Wiener, *Physical Theory*, pp. 264 ff. It would be absurd, however, to conclude that Duhem was not aware of the significance of German physics. His reviews show the reverse. In his review of Kirchhoff, *Vorlesungen ueber mathematische Physik*. Vol. 2. *Optik* (Leipzig, 1899), Duhem lamented that the first volume on mechanics, in its third edition in Germany, "le guide indispensable de quiconque veut étudier la physique mathématique," had not been translated into French. *Revue des questions scientifiques* 32 (1892): 273–75. In his review of the fourth volume, *Theorie der Wärme* (Leipzig, 1894), Duhem noted, "le souci de l'élégance et de la concision, souci parfois excessif, peut-être." *Revue des questions scientifiques* 36 (1894): 266–67. See ibid., 33 (1893): 264–65, for a review of the third volume, *Electricität und Magnetismus*.

Duhem's areas of expertise: electrical theory. The difficult theory of electricity and magnetism was built on principles enunciated with Gallic clarity by Poisson and Ampère, who served as guides to the most famous German physicists—Karl Friedrich Gauss, Wilhelm Eduard Weber, and Franz Ernst Neumann, who, inspired by the *esprit de finesse* and disciplined by the *esprit géométrique*, built one of the most powerful and harmonious doctrines of physics. But, lamented Duhem, the doctrine had recently been completely upset by the geometric minds of the Germans. Of course, the beginning of the attack came from Maxwell, whose work, he hastened to point out, was not Germanic; indeed, Maxwell's *esprit de finesse* was the most spontaneous since that of Fresnel. In Maxwell, then, Duhem found a most desirable combination of the two types of mentality needed for great scientific work. He co-opted Maxwell into the pantheon of geniuses whose work was exempt from the vices that pervert the science in which one mentality or the other predominates. Maxwell had become nearly French. This was a change from Duhem's estimate of Maxwell in 1893. The continuation of Maxwell's work by Helmholtz showed the same desirable combination of mentalities. But Helmholtz' theory, so satisfying for any harmoniously constituted reason, did not find favor in Germany. Hertz, Helmholtz' pupil, gave to Maxwell's thought a form from which the *esprit géométrique* had rigorously excluded the *esprit de finesse* by constructing a system that accepted Maxwell's equations as axioms or orders. The use of a doctrine that denied the existence of permanent magnets while making use of them in experiments was another illustration of the inability of the geometric mind to apply its deductions to the data of experiment.[41]

Duhem's *Angst* over the imperialism of the geometric mind in Germany is revealed best in his comments on developments in atomic physics in his time. Although it is evident that he was still reluctant to change his paradigms,[42] solidly rooted in classical physics, he no longer says that the new atomic physicists are uncles of Alladin exchanging new lamps for old, which had been his atti-

41. For some quasi-technical comments, see Duhem, *Revue des deux mondes,* pp. 676–77.

42. A typical attitude, even on the part of the innovators: Planck refers to a striking and unpleasant surprise at the turn of the century when classical physics found itself impotent to make advances in heat radiation, light rays, and electromechanics. Max Planck, *The Philosophy of Physics* (New York, 1936, 1963), p. 18.

tude in the 1890s. He warned that to establish the electrodynamics of the putative electron, one should imitate the prudent method by which Ampère, Weber, and Neumann had formed the electrodynamics of a conducting body: delicate experiments, penetrating intuitions, and arduous discussions, such as were begun by Weber, Riemann, and Clausius, rather than by the quicker and less painful method of algebraism. Rallying to the hypothesis of Lorentz, German scientists deduced the physics of the electron. Duhem's alarm was based on his claim that this physics was built on a generalization of Maxwell's equations as the fundamental axiom; this was to build on a worm-eaten foundation because Maxwell's equations had not been cleansed of their basic vices. Worse, rational mechanics was being shaken in its very foundations by this new physics, which proposed abandoning the principle of inertia and entirely transforming the idea of mass. Such was the devastation of the geometric mind in sweeping through classical physics, although it used the theories of rational mechanics to interpret the readings of the instruments providing its information.[43] Duhem should not be taken to represent French science's reaction against the new physics. Poincaré's response concerning quantum theory, for example, was quite different: "The beginnings of the challenge of the quantum theory to the foundations of physics coincided with the career of Henri Poincaré. One of the last masters of classical science, Poincaré became for a brief time one of the few of the new science."[44]

Poincaré's reaction to relativity was, however, quite different. Although it is "remarkable . . . that he should have stopped short of the full presentation of relativity theory as we know it," he "remained unshakable against Einstein's interpretation to the end." Here he was not, in Holton's lingo, "themata-prone" but remained the "brilliant conservator of his day." This role was not unlike that of Duhem: he was also hostile toward relativity theory, which he regarded as a scientific *credo quia absurdum*. But Holton notes the reversal of roles on the topic of quantum mechanics when Poincaré and Einstein met in 1911. The conservation element in French science is probably no more indicative of the whole situation than the innovation aspect is of German science. Many of the attitudes of Poincaré and Duhem were not part of the intellectual baggage

43. Duhem, *Revue des deux mondes*, pp. 674–78.
44. Russell McCormmach, "Henri Poincaré and the Quantum Theory," *Isis* 58, no. 191 (1967): 37–55. Poincaré died in 1912.

of the "innovators"—Langevin, Perrin, Nordmann, de Broglie—who were critical of the masters' "conventionalist epistemology" and kept France in the vanguard of scientific developments in this "time of turmoil in the physical sciences as well as in the philosophy of science."[45] Duhem complained that the new physics conflicted not only with other physical theories but with common sense. In order to make the equations of the new theory agree with Michelson's experiment,[46] the geometric mind of the German physicists overthrew the common-sense ideas on space and time by uniting them through an axiom that was really an algebraic definition of time— the principle of relativity. Duhem also criticized the idea that the speed of no object could exceed the speed of light; to Duhem this was a logical impossibility, not a physical one.[47] Duhem thought that the geometric minds of the German physicists took great delight in sweeping away the old doctrines based on observation and experiment and in constructing a complete physics based on the principle of relativity. Proud of its algebraic rigidity, the physics of relativity constructed by the geometric mind scorned the good sense that falls to the lot of all men.[48] Duhem was shocked by the Einsteinian idea

45. Gerald Holton, "On the Thematic Analysis of Science: The Case of Poincaré and Relativity," in *Mélanges Alexandre Koyré* (Paris, 1964), 2: 257–68; "Mach, Einstein, and the Search for Reality," *Daedalus* 97, no. 2 (1968): 636–73; "On the Origins of the Special Theory of Relativity," *American Journal of Physics* 28, nos. 1–9 (1960), partly reprinted in L. Pearce Williams, ed., *Relativity Theory: Its Origins & Impact on Modern Thought* (New York, 1968), pp. 100–107. See also the selection from Whittaker, in Williams, pp. 94–99, especially p. 99. For comment on the limitations of "phenomenalistic positivism," see Holton, in *Daedalus* 97 (1968): 656, and Mary B. Hesse, *Forces and Fields. The Concept of Action at a Distance in the History of Physics* (London, 1961).
46. A conventional myth in physics textbooks according to Michael Polanyi, *Personal Knowledge.* See pp. 9–15 for a comment on the nonrole of the Michelson-Morley experiment in the formation of Einstein's theory and on the fact that the experiment does not agree with the theory. But see Adolf Grünbaum's challenge to this interpretation in "The Genesis of the Special Theory of Relativity," reprinted in Williams, pp. 108–14, especially pp. 113–14.
47. This is presumably the conclusion of contemporary physicist Gerald Feinberg, whose imaginary particle—the "tachyon"—travels faster than light but is as yet undetected by researchers. See *Time*, February 14, 1969. The Asimov-Clarke debate in *Fantasy and Science Fiction* provides a nontechnical insight into the solar claustrophobia inspiring Feinberg.
48. Duhem, *Revue des deux mondes*, pp. 680–83. Emile Meyerson, *Identité et réalité* (Paris, 1908), ch. 11: "Le sens commun," pp. 328–52, makes substantial use of Duhem. Duhem's position should not be dismissed as anti-Germanism. For examples of similar scientific reaction to Einstein, see Williams, as well as Polanyi, *Personal Knowledge*, pp. 9–15. Some French scientists

that "There is . . . no logical way leading to the establishment of a theory but only groping constructive attempts controlled by careful consideration of factual knowledge."[49] "A theory as noncommonsensical as Einstein's" has to be partly explained in terms of Einstein's "taste for 'inner perfection' in science," which does not proceed according to Duhemian logic but according to "an aesthetic which logicians of science have not yet reduced to empirical terms, or to intersubjective agreement."[50]

The immense advantage that the geometric mind could enjoy in developing a science whose fundamental principles were already laid was evident in chemistry, which had undergone a prodigious development in nineteenth-century Germany. Duhem argued that the role of the Germans was still small when J.-B. Dumas, followed by Laurent, Gerhardt and Wurtz (both from Strasbourg), and the English Williamson, carried out researches that through the theory of types made up modern organic chemistry. Only the German Hofmann, the rival of Wurtz in the discovery of amines, could be regarded as comparably important. But these researches introduced into chemistry the language and procedures of geometry. And the German Kekulé formulated the rules accurately and systematically. The take-off of German chemistry coincided with the acceptance of atomic notation; this, with the help of the rules furnished by the part of algebra called *Analysis situs* (topology), permitted prediction, enumeration, and classification of reactions, syntheses, and isomers of carbon compounds. So organic chemistry, subject to the geometric mind, brought forth from numerous German laboratories thousands and thousands of new organic compounds, all classified and described according to principles drawn from topology.[51] The subtle mind then had far less to do in organic

---

accepted relativity. See the technical work of E. M. Lémeray, *Le Principe de relativité* (Paris, 1916): "Cours libre professé à la Faculté des Sciences de Marseille pendant le premier trimestre 1915." Lémeray accepted relativity as "un instrument de coordination et, par conséquent, de découverte d'une puissance singulière." H. Dingle has been arguing against special relativity throughout the 1960s: see "The Case Against Special Relativity," *Nature* 216 (1967): 119–22, a restatement of his earlier case in *Nature* 195 (1962), which brought a reply from Max Born. W. H. McCrea, "Why the Special Theory of Relativity is Correct," *Nature* 216 (1967): 122–24, defends Einsteinian orthodoxy. A correspondence over the issues begins in *Nature* 219 (1968): 790–93.

49. Einstein, quoted in Holton, *Isis* 60 (1969): 156.

50. C. C. Gillispie, *The Edge of Objectivity* (Princeton, 1960), quoted in Holton, *Isis* 60 (1969): 180.

51. In the Kekulé memorial lecture, F. R. Japp said that three-fourths of

chemistry, while the geometric mind became more indispensable every day.[52]

In spite of his criticisms of German science, Duhem recognized its genius. In concluding his article, he noted that both French and German science deviated from the ideal and perfect science, although they deviated from it in opposite directions. In order for human science to develop fully and exist in a harmonious equilibrium, German and French science would have to co-exist, without trying to supplant one another: "Each must understand that it finds in the other its indispensable complement." In 1894 Duhem had recommended Neumann as a patron saint for young French physicists:

> Il me semble que le doyen des physiciens, l'illustre F. E. Neumann, doit contempler avec un légitime orgueil cet admirable ensemble de leçons embrassant toutes les parties de la physique, qu'il a professées il y a vingt ans, et qui, publiées aujourd'hui par ses élèves demeurent des modèles que n'égale presque aucun des enseignements de physique de l'Europe. Quelle étendue de connaissances! quelle pénétration de toutes les théories! et en même temps, quelle clarté, quelle élégance dans l'exposition! Les sept volumes qui renferme les cours de M. F. E. Neumann devraient être dans la bibliothèque de tous les jeunes professeurs de physique, auxquels ils fourniraient avec une égale liberalité le fond d'un enseignement solide, et une forme parfaite.[53]

In studying German science, the French would find solid proof of truths they had discovered and formulated without complete certainty, or else they would find the refutation of errors they had committed under the intoxication of an imprudent intuition. It was equally useful for the Germans to study French science: there they would find problems for their patient analysis to solve as well as

---

modern organic chemistry is directly or indirectly derived from Kekulé's "ring" theory of the constitution of benzene. *Encyclopedia Britannica* 13: 315.

52. Duhem, *La science allemande*, pp. 37–42; *Revue des deux mondes*, 683–84. He made the same argument for physical chemistry and stereochemistry, but noted that in mineral chemistry, where atomic notation had only a restricted usage, the *esprit de finesse* was still the instrument for untangling the complexity of reactions and for classifying compounds.

53. Duhem, review of F. Neumann, *Vorlesungen über mathematische Physik gehalten an der Universität Königsberg. Siebentes Heft. Vorlesungen über die Theorie der Capillarität* (Leipzig, 1894), *Revue des questions scientifiques* 36 (1894): 267.

the protests of common sense against the excesses of their geometric minds.

Henry Guerlac argues that "there are subtle cultural tendencies which set apart the scientific achievements of every nation. The French animus in favor of pure science is just a tendency. . . ." Guerlac sees France's achievements as more theoretical and less empirical, more mathematical and less experimental, than England's. So "the typical French achievement . . . seems . . . rationalistic, one of synthesis or great theoretical insight: such for example as Laplace's . . . *Mécanique celeste*; Carnot's pioneer speculations into the nature of heat and energy . . . Lavoisier's reform of chemistry. Despite the weighty instances of Claude Bernard and Pasteur, it is the Anglo-Saxons, not the French, who seem to have produced the mainly experimental men, the Franklins, the Faradays, and the Joules."[54] This is an obvious exaggeration, but such is the danger in the Procrustean procedure of trying to judge the national characteristics of science. Duhem's more subtle analysis avoids some of the dangers of oversimplification but is laden with other equally serious limitations. Duhem still viewed German science in the nineteenth century as derivative of the work of the great French thinkers. It could not be otherwise on the basis of his analysis: intuition, the prerogative of the *esprit de finesse*, discovers truth; demonstration, the area of the *esprit géométrique*, comes after; it was the relationship of the architect to the mason. The geometric mind in inspiring German science gave it the force of a perfect discipline, but this method could only lead to disastrous results if it continued under the orders of an arbitrary and mad algebraic imperialism. If German science were to escape this fate and produce useful and fine work, it would have to be put under the orders of the chief depository of common sense in the world, French science: *Scientia germanica ancilla scientiae gallicae.*[55] The sorcerer's apprentice must be saved from his own magic.

Duhem did not always remain at the high level of detachment

54. Guerlac, in Earle, pp. 82–83.
55. Duhem, *Revue des deux mondes*, pp. 685–87. See also Duhem, *La science allemande*, pp. 88–99. But in this more patriotic work Duhem lamented the servile copying of German science by French science, and like Picard, called for its liberation: after victory there would be the task of giving back to the Fatherland its fullness and purity of soul. See also René Lote, *Les origines mystiques de la science "allemande"* (Paris, 1913), and *Les leçons intellectuelles de la guerre* (Paris, 1917).

indicated by his initial comments on German science. To have done so would have been against the nature of the indiscreetly frank physicist who had sacrificed for the Catholic cause a goodly part of the emoluments and upward mobility offered those in the Third Republic's educational establishment who supported the secular republican ideology.[56] In *La science allemande* this conservative "monarchist" and sympathizer of Maurras became as virulently patriotic as the best republican. *La science allemande* consisted of four lectures given at Bordeaux under the auspices of the Association des Etudiants catholiques de l'Université in February and March, 1915. The lectures were dedicated to the Catholic students of the University of Bordeaux, who had requested them, with the hope that with the help of God the lectures would maintain and promote in the students and their comrades the "clear genius of our France." Duhem agreed with the abbé Bergereau that the invasion of the soil of France was only one aspect of the conflict; foreign thought had enslaved French thought. Duhem regarded his intellectual task as combat duty, which, although without danger or glory, he would fulfill with complete devotion, taking his humble role in the national defense.[57]

56. For details, see my forthcoming essay in *The Catholic Historical Review*.
57. Duhem, *La science allemande*, pp. 3–4.

4. Postwar Ostracism of the Sorcerer's Apprentice

The savage attacks upon German science made by French scientists during World War I did not represent an entirely new phenomenon in France. In many cases the attacks were a continuation and culmination of an anti-German movement going back at least to 1870.[1] The debate was seen clearly in chemistry and was not restricted to contemporary developments but involved a reexamination of the history of chemistry and a reassessment of the relative roles played by the great French and German chemists in its development. In 1874 Wurtz published the first volume of the *Dictionnaire de chimie pure et appliquée*, in which he declared that "Chemistry is a French science." After it existed for centuries as a collection of obscure and often incorrect recipes, Stahl tried at the beginning of the eighteenth century to give it a scientific basis, but he failed—this glory was Lavoisier's. But Ostwald challenged this contention, describing Lavoisier's progress in his theory of oxidation over Stahl's phlogiston theory as generally exaggerated in the history of science. Stahl had done the essential, the systematization of combustions; the only thing to do after Stahl was to take symmetrically the inverse of the ideas relative to combination and decomposition.[2] In 1915 Duhem came to the rescue of Lavoisier

1. Cf. Amédée Latour: "C'est grâce à notre esprit affolé de propagande, que la science tudesque s'est répandue dans le monde savant tout entier." Cited by Gaucher, "La thérapeutique commerciale des Allemands," in Petit, p. 154.
2. *L'évolution de la science, la chimie* (Paris, 1909). Translated into French by Marcel Dufour. The development of a hagiography of German scientists in contemporary German works led to bitter French complaints about the slighting or ignoring of their own scientific heroes. Picard, in Petit, pp. 290–91, complained of the neglect of Lavoisier, Sainte-Claire Deville, Berthelot, and Pasteur in German works in favor of counterheroes like Stahl, Wöhler, and Koch. Armand Gautier, "La science et l'esprit allemands," in Petit, p. 117, complained about Ehrlich's citing only one French name in his famous pam-

with a short history of chemistry, admitting only that Stahl had prepared the way for Lavoisier.[3] Stahl was really only worthy to unloose the latchet of Lavoisier's shoes; the messiah of chemistry had been French. In the twentieth century this quarrel did not die but continued and expanded to other areas as the German challenge in science grew stronger.[4]

It is clear that the French writers frequently admitted German superiority in areas that could be measured, claiming that French superiority could not be measured. Richet adopted a view easier to defend historically, at least until the last quarter of the nineteenth century, namely, that the French had often made key discoveries. Many of the French writers had an obsession with the history of science, which gave them a counterbalance to contemporary overshadowing by Germany. Except for the historically based and conceptually organized propaganda of Duhem, the efforts were for the most part elegant propaganda catalogs.

Few were as honest as Boutroux, who admitted that Germany enjoyed world scientific leadership. This had been a conventional judgment in France before World War I. Duhem had denounced Berthelot for obstructing the growth of the new thermochemistry in France and allowing leadership to pass to others, including Germany.[5] Sainte-Claire Deville had interpreted the defeat of 1870–71 as France's having been vanquished by science. Pasteur had at the same time regretted the loss of the scientific leadership France had

---

phlet on the therapeutic application of arsenic, whereas in 1902, six years before Ehrlich's pamphlet, Gautier had made a full report to the academies.

3. Pierre Duhem, *La chimie est-elle une science française?* (Paris, 1916). See Henry E. Guerlac, "Some French Antecedents of the Chemical Revolution," *Chymia* (Annual Studies in the History of Chemistry) (1958), pp. 73–112, and *Lavoisier—The Crucial Year. The Background and Origin of His First Experiments on Combustion in 1772* (Ithaca, 1961).

4. The attack was frequently aimed at the prostitution of *Isis* for commercial purposes. See, e.g., E. Gaucher, "Le '606' et les médecins," *Journal de médecine interne*, November 11, 1911, and the *Annales des maladies vénériennes,* December 1911. In these works the claim was made that German therapeutic novelties were a money-making device. Gaucher, in Petit, pp. 154–66, returned to the attack. He blamed French doctors for their ignorance of the fraudulent nature of remedies supplied by German commercial interests. His chief examples were the treatment of TB by Koch's inoculation and of syphilis by Ehrlich's dangerous arsenic preparation, "606." See also Edmond Perrier, "L'Allemagne savante," in Petit, pp. 275–82.

5. Duhem, "Thermochimie à propos d'un livre récent de M. Marcellin Berthelot," *Revue des questions scientifiques*, IIe sér. 21 (October 1897): 361–92.

enjoyed in the early nineteenth century. Dumas had declared that the future belonged to science.[6]

In spite of this attitude during the Third Republic, when scientists like Berthelot and Painlevé[7] played an important political role (which explains to a considerable extent the importance of science in the Republican ideology), the post–1870 outlook did not prevail during World War I. In England, however, the war provided an opportunity for scientists to draw attention to the "national neglect of science." In the Romanes lecture of 1914, E. B. Poulton, Hope Professor of Zoology and president of the Linnean Society of London, delivered a savage attack on the predominance of classics in education, "the dullness and stupidity of the Foreign Office," and the domination in Parliament and especially in the government of "the spirit that is most antagonistic to science—the spirit of the advocate." He contrasted this outlook, which explained such tragedies as the transfer of the coal-tar industry to Germany, itself a result of the lack of government support for chemical research, with the "patriotic position taken for many years by . . . Nature," whose warnings, if heeded, would have ensured long ago "Germany's defeat and the freedom of the world. . . ."[8] Basically, of course, England's problem of adjusting to the new scientific pre-eminence of Germany was the same as that faced by France, but because of the different internal conditions and different experiences with Germany, each country's reactions had its distinctive characteristics and did not reach the same stages of development at the same time. It is also probable that the expansion of science under the Third Republic along with the prewar signs of "renewal" in French science, especially in physics with the Curies, Perrin, and Langevin, gave the French scientist a feeling that France was regaining at least part of her lost scientific glory, which had been so tarnished in 1871.

The French mind had been in a state of shock. So long the scientific leader of Europe,[9] the trauma of being overtaken, especially

6. René Vallery-Radot, *The Life of Pasteur* (London, 1906; New York, 1960), pp. 195–96, 251.

7. See Paul Painlevé, *De la science à la défense nationale. Discours et fragments.* Préambule de M. Jean Perrin (Paris, 1931).

8. E. B. Poulton, *Science and the Great War* (Oxford, 1915), pp. 4–9. On p. 8 he quotes Perkin's attack on English universities and schools and his remark that Germany had ten times as many advanced research students as England.

9. "The period of French scientific supremacy extends from about 1750

by its chief rival on the continent, required nearly half a century to be absorbed by the national conscience. France was the victim of the hubris arising from its pre-eminent position in the Enlightenment, a position enhanced by its material conquests in the Revolutionary and Napoleonic periods. This was the same intellectual egotism that for more than half a century myopically regarded Kant as a neo-scholastic. The war provided an excellent opportunity for the French to carry out a reconsideration and point out all the vices of German science, especially its responsibility for the war. Here was an opportunity to show that French science had kept to the path of virtue, while the Faustian *Sehnsucht* of German science had damned it irretrievably. It was irrelevant that most of the same vices were endemic in France.[10] It was difficult for the French to adjust their image of French science, an image correct for the late eighteenth and early nineteenth centuries, to the reality of the late nineteenth and early twentieth centuries. To indulge in an activity of the *esprit géométrique,* we may say that Boulding's general judgment on the vitiating danger of image persistence applies aptly to the French and their science: "It is not fanciful, however, to detect pathological relationships at certain times and places in history between the public images and the rest of the social universe. Curiously enough, it is often the most successful images that become the most dangerous. The image becomes institutionalized in the

---

through the 1830s and 1840s. . . . During this long hegemony, French leadership extended into every field of natural science, and Paris was sought out by students of chemistry, mathematics, natural history, and medicine from other European countries and from America." Guerlac, in Earle, p. 83.

10. Pasteur had unsuccessfully campaigned for the Senate in 1876 using the slogan "Science et Patrie"; he emphasized that France had been victorious in 1792 because "Science had given to our fathers the material means of fighting." Vallery-Radot, *Pasteur,* p. 248. Pasteur's approach, in the tradition of French science in the Revolution, is the one that probably emerged triumphant from the war; Germany was to be emulated, not forgotten. Charles Nordmann, "L'industrie chimique française et la guerre," *Revue des deux mondes* 38 (April 1, 1917): 697–708, lamented the separation of industry and science in France compared with their symbiosis in other countries. His arguments are in sharp opposition to the Duhemian idea of keeping French science free from the vices of English and German science. Nordmann brings Danton up to date: "Pour vaincre les ennemis de la Patrie que faut-il? De l'acide sulfurique, encore de l'acide sulfurique, toujours de l'acide sulfurique!" The French must organize their chemical industry on the American and German models. See also Charles Moureu, *La chimie et la guerre; science et avenir* (Paris, 1920). On the German side, see Richard Anschütz, *Die Bedeutung der Chemie für den Weltkrieg* (Bonn, 1915).

ceremonial and coercive institutions of society. It acquires there a spurious stability. As the world moves on, the image does not."[11]

This became brutally clear in the interwar period, when France could exploit Germany's defeat to institutionalize in an ostensibly international organization the image she had created of German science. Picard made the radical proposal that all intellectual relations with Germany should be ended as soon as possible. There should be no return to the remarkable period of international scientific cooperation prevalent between the Franco-Prussian War and World War I.[12] Only by a ruthless ostracism of Germany in the "international" scientific community could the nefarious effects of German thought be stopped. Published works should be read, but personal contact with German scientists should cease. It was especially important to ban Germany from international congresses, which, in Picard's opinion, she had exploited as a springboard for the conversion of others to "Germanism."[13] Brigitte Schröder has shown how the split that had developed in the scientific community by 1914 continued after the war. The "Central Powers" were excluded from the International Research Council founded in 1919. German was pointedly refused status as an official language of the organization. Even after the thaw following 1925, difficulties persisted between Germany and France, in contrast to a developing collaboration among Germany, the United States, and England. German scientists remained so hostile to the organization that in 1926 they ignored an invitation to join. While the political world was moved by the Locarno spirit, the Franco-German scientific world remained anchored to the position created during the war.[14]

11. Kenneth E. Boulding, *The Image* (Ann Arbor, Michigan, 1956), p. 79. It is possible to consider French scientists' opinion and use of German science within the framework of "the scientific reception system" as set forth by Alfred de Grazia, *The Velikovsky Affair: Scientism vs. Science* (New Hyde Park, N.Y., 1966). But any model concocted would involve a juggling of the variables of the rational, the dogmatic, and the power models, if one wishes to approach the distorted reality of the wartime France.

12. See Brigitte Schröder, "Caractéristiques des relations scientifiques internationales, 1870–1914," *Journal of World History* 10, no. 1 (1966): 161–77.

13. Picard, in Petit, pp. 297–99; Dastre, Delbet, and especially Boule, passim, make the same points.

14. Brigitte Schröder, "Science and Politics," paper read at the History of Science meeting in Washington, D.C., December 1969. See also Brigitte Schröder, "Les professeurs allemands et la politique du rapprochement," *Annales d'études internationales*, no. 1 (1970): 1–22; "La conférence interalliée des académies scientifiques à Londres" and "La deuxième session de la conférence interalliée des académies scientifiques," *Revue scientifique*, nos. 16–23

The general basis of much French criticism of German science was that science and technology are two separate although related activities, an idea generally accepted by modern scholarship. Kuhn has advanced the idea that "a thorough historical survey would show that traditionalistic, abstract, and pure scientific thought tend to flourish together and that innovating, concrete, and applied scientific thought also show some statistical clustering. . . ." This can be seen in the "contrast between Greece and Rome in Antiquity, France and Britain from 1750 to 1850, and France and the U.S. throughout most of our history."[15] For Kuhn "the scientist and the inventor are often profoundly different types," and "they flourish under rather different cultural and social circumstances."[16] At least until the end of the nineteenth century science and technology seem to be separate areas of activity and enterprise. Kuhn and Kaplan summed up the issue for the 1959 Utah Research Conference: there is "overwhelming historical evidence that a cultural and institutional climate highly favorable to eminence in technology need not be favorable to research in the more abstract or more fundamental sciences." There is also "good reason to suppose that the environmental requisites for creativity may vary significantly from science to science." Unfortunately for the propaganda case made by French scientists in World War I, "Germany during the half

---

(Nov. 1918): 673–76; nos. 4–11 (Jan. 1919): 21–23; and "La conférence interalliée de la chimie," ibid. (July 5, 1919), pp. 399–403. The entire issue of *Chimie et industrie*, May 1919, was devoted to the conference on chemistry. One of the strongest anti-German statements came from Eugenio Rignano, editor of the Italian *Scientia*, who wrote to *Nature* (97 [Jan. 25, 1917]: 408–9): ". . . the numberless *Archive*, *Jahrbücher*, *Zeitschriften*, *Zentralblätter*, and so on . . . have gradually monopolized the whole of the scientific production of the world. Thus were apparently built up international scientific organs, but in reality German instruments of control and monopoly of science. . . . [The] pacific war of liberation from Germany hegemony . . . must be continued . . . after peace comes, [and] must also be carried into the scientific domain." Cited in Schröder, *Annales d'études internationales*, no. 1 (1970): 9n24.

15. Kuhn, *Utah Research Conference*, p. 177.

16. Kuhn, "Comment on Scientific Discovery and the Rate of Invention," in *The Rate and Direction of Inventive Activity: Economic and Social Factors*, A Report of the National Bureau of Economic Research (Princeton, N.J., 1962). Kuhn finds corroboration of his argument in Donald W. Mackinnon's paper "Intellect and Motive in Scientific Inventors: Implications for Supply," ibid., pp. 361–78. See also R. P. Multhauf, "The Scientist and the 'Improver' of Technology," *Technology and Culture* 1, no. 1 (1959): 38–47, and Derek J. de Solla Price, "Is Technology Independent of Science?" ibid. 6, no. 4 (1965): 553–68. The latter issue of *Technology and Culture* is devoted to "the historical relations of science and technology."

century after 1870 provides perhaps the only historical example of simultaneous pre-eminence in both sorts of work."[17]

There was a survival well into the twentieth century, in a limited literary area at least, of the Bergson-Boutroux strain of antimechanistic thinking in the French reaction to the menace of subjugation by the Teutonic *Kultur*. Georges Bernanos' *La France contre les robots*[18] is a superb example: France is the last holdout of the Erewhonians. Bernanos made a point similar to one that the critics of German science harped on—the civilization of the machine is in no way the work of scientists, but rather of greedy men who have created it unintentionally as the necessities of their business required it: "Un monde gagné pour la Technique est perdu pour la Liberté." The paradise of the civilization of the machines would be opened by the two magic words obedience and irresponsibility. But French civilization, on whom had fallen the mantle of Hellenic civilization, had worked for centuries to form free men, men fully responsible for their acts; France would therefore refuse to enter the paradise of the robots, a mirage leading to war and destruction.[19] Jules Simon had made a similar point much more mildly in 1885: "L'avenir n'est pas dans le développement excessif de l'érudition mais dans la culture du talent. Il est peu désirable de copier en cela les allemands qui peuvent à peine arriver à comprendre les qualités intellectuelles françaises."[20] What Bernanos attributed to the domination of the civilization of the machine, in both democratic and totalitarian nations, was seen by the typical French scientist, looking at Germany in World War I, as the inevitable consequence of Germany's political, social, and scientific development. Mallarmé,

17. Kuhn and Norman Kaplan, "Environmental Conditions Affecting Creativity," *Utah Research Conference*, p. 314. See also Irving H. Siegel, "Scientific Discovery, Invention, and the Cultural Environment," in *The Patent, Trademark, and Copyright Journal of Research and Education* (Fall, 1960), summarized as "Scientific Discovery and the Rate of Invention," in *The Rate and Direction of Inventive Activity*, pp. 441–50.

18. Finished in exile in 1944, published in Paris in 1947, and recently brought out in a new edition, which testifies to a revival of Bernanos now occurring in France, even at the *Le Figaro littéraire* level.

19. *La France contre les robots*, pp. 26, 189, 219–20. Bernanos noted that the civilization of the machine would have no need of the French language, a work of art, the flower and fruit of a civilization absolutely different from that of the machine, p. 178. See the interesting diatribe of Paul Goodman, "Can Technology be Humane?" *The New York Review of Books*, November 20, 1969, pp. 27–34.

20. *Conseil supérieur de l'instruction publique*, July 25, 1885, AN, F17 12967.

drawing upon the poet's capacity for economy of thought, had dealt with the universal issue in his *Sonnet à la science.*[21] The French scientist in World War I limited its application to Germany. This was a delusion produced by the war.

Claude Digeon concluded, as a result of his study *La crise allemande de la pensée française (1870–1914)*, that between 1870 and 1914 the "German question" provoked and sustained a long crisis in French thought. The crisis was launched by the war of 1870 and perpetuated by the historical and intellectual consequences of that war. Overlooking the nearly tropistic use of the *crise* cliché, put forward all too often as a substitute for explanation in modern French history, we must keep in mind that the so-called crisis in French thought, which might better be called an increasing awareness of intellectual developments in Germany, began much earlier than 1870.[22] This is certainly the case in science, as illustrated by the numerous discussions, reports, and complaints that issued from the scientific community after 1830. Digeon argued that in the history of French images of nineteenth-century Germany the war of 1870 marked the end of the period dominated by the "tradition staëlienne." Within the scientific community it is difficult to see that the war of 1870 really induced a mutation in French scientific thought about the nature of German science, whatever may have been the new popular awareness of German science and technology. The international nature of the scientific enterprise in the nineteenth century precluded any surprises except in the popular mind. It was the development of German science and technology themselves that made a serious impact on the French scientific mind. This is not to deny that the war of 1870, like that of 1914, caused the French scientist to be more keenly aware of the nature of the German achievement. The hand of the French scientist also seems to have

21. "Science, tu es la vraie fille du vieux temps, qui changes toutes choses par ton oeil scrutateur. Pourquoi fais-tu ta proie ainsi, du coeur du poëte, Vautour dont les ailes sont de ternes réalités? Comment t'aimerait-il? ou te jugerait-il sage, toi qui ne le laisserais point, dans la promenade de son vol, chercher un trésor en les cieux pleins de joyaux, encore qu'il y soit monté d'une aile indomptée. N'as-tu pas arraché Diane à son char? et chasse du bois l'Hamadryade qui cherche un refuge dans quelque plus heureux astre? N'as-tu pas banni de son flot la Naïade, du vert gazon l'Elfe et moi des rêves d'été sous le tamarin?" Stéphane Mallarmé, *Oeuvres complètes* (Paris, 1945), p. 220.

22. Digeon, *La crise allemande*, p. 535; Paul, "In Quest of Kerygma: Catholic Intellectual Life in Nineteenth-Century France," *The American Historical Review* 75, no. 2 (1969): 387–423.

been strengthened in his attempt to get government support for science and technology on an unprecedented scale. But in spite of complaints about material deficiencies, we do not find in French science the national inferiority complex emphasized in literature and in the myth of *revanche*.

As we saw, World War I encouraged and even demanded the application of science to industry, especially in emulation of Germany. This point of view was put forward in wartime rhetoric by Painlevé in 1915, as Président de la commission de la marine:

L'avantage que l'Allemagne conserve encore, ce n'est pas, comme on le croit souvent, sa préparation préméditée à la guerre; le bénéfice en est épuisé après un an de batailles; c'est sa puissance de production dans les industries métallurgiques et chimiques. Ses fabriques se sont trouvées tout organisées pour la guerre; il lui a suffi de leur faire donner leur plein. Sa natalité lui a permis en même temps de laisser, là où il fallait, ses savants, ses ingénieurs, ses ouvriers.

La France, elle, toute meurtrie des coups de l'ennemi, a dû décupler sa métallurgie, créer des industries chimiques inexistantes. Malgré l'invasion, malgré une mobilisation automatique qui vidait, dès les premiers jours, ses ateliers, ses arsenaux, ses laboratoires, elle a fait des prodiges. Cela ne suffit pas; il faut qu'elle se surpasse encore.[23]

But it is an exaggeration to state that "French attempts to modernize science date from the conclusion of World War I."[24] The roots of this movement lie deep in the nineteenth century. Indeed, the movement has always been strongly represented in French science itself. By the beginning of the twentieth century the movement was even strongly entrenched in the Sorbonne. In the provincial universities there had always been a strong emphasis on local industry and agriculture.[25] That this trend toward an alliance between industry and science had firm opponents, the defenders of "ivory tower" science, should not obscure the fact that by the last decade of the nineteenth century it was clear that they had lost the battle

23. Painlevé, *De la science à la défense nationale*, p. 21. In October 1915, Painlevé became Ministre des inventions and a member of the Comité de guerre; in 1917, he became Minister of War, a position he held again in 1925 and 1925–29; and in December 1930 and January 1931, he was Ministre de l'air. He was also the Président du conseil in 1917 and in 1925.

24. Guerlac, in Earle, p. 97.

25. See Paul, *French Historical Studies*, forthcoming.

even in the Sorbonne itself. It is not easy to determine the role of the French scientist's consciousness of scientific developments in Germany in this transformation. Certainly, in spite of the internal impetus within French science toward the same goals as science in Germany, the example of developments in Germany had been a constant preoccupation of the French scientific mind from before the days of the Second Empire. The German example often provided models to emulate and sometimes to avoid. Faced with the rise of British and German science and technology, the French had to adjust themselves to the painful fact that they were no longer the scientific pedagogues of the world but equal co-workers, sometimes pioneers but more frequently followers, in the great international enterprise that science has become in modern times. But this was achieved only after nearly a century of rethinking and the catalyst of two devastating wars. Even during the heat of battle, men of reason and intelligence recognized the changes that had occurred since the early nineteenth century and the more modest role that France consequently occupied in the scientific firmament:

A travers les siècles, la Science française a su conserver les caractères distinctifs de son génie et elle fut fidèle à son idéal. Il ne faut pas croire cependant que ce respect pour ses traditions, cet attachement à ses habitudes aient gêné sa marche en avant et paralysé ses progrès. Elle a bien souvent, au contraire, montré une extraordinaire facilité d'adaptation et une souplesse parfaite. Sur le terrain scientifique, comme sur d'autres, la France a été la plus révolutionnaire des nations; elle a brisé des cadres anciens, institué des régimes nouveaux et, sans préjugé, sans parti pris, elle s'est solidement installé sur les positions conquises.

. . . . .

Mais, sous des formes entièrement nouvelles, elle [la Science française] garde de sa tradition une partie immatérielle qui n'est pas un fâcheux reste du passé . . . elle ne prétend pas être la seule de par le monde, elle sait seulement qu'elle a toujours eu et qu'elle conserve une très grande place et, généreuse et hardie selon sa coutume, elle a, sans arrière-pensée d'imposer sa domination, la volonté d'être parmi les premières dans la marche triomphale de l'esprit humain vers la Vérité.[26]

26. Lucien Poincaré, "Préface de la première édition," *La Science française* 1: x–xi.

UNIVERSITY OF FLORIDA MONOGRAPHS

Social Sciences

1. *The Whigs of Florida, 1845–1854,* by Herbert J. Doherty, Jr.

2. *Austrian Catholics and the Social Question, 1918–1933,* by Alfred Diamant

3. *The Siege of St. Augustine in 1702,* by Charles W. Arnade

4. *New Light on Early and Medieval Japanese Historiography,* by John A. Harrison

5. *The Swiss Press and Foreign Affairs in World War II,* by Frederick H. Hartmann

6. *The American Militia: Decade of Decision, 1789–1800,* by John K. Mahon

7. *The Foundation of Jacques Maritain's Political Philosophy,* by Hwa Yol Jung

8. *Latin American Population Studies,* by T. Lynn Smith

9. *Jacksonian Democracy on the Florida Frontier,* by Arthur W. Thompson

10. *Holman Versus Hughes: Extension of Australian Commonwealth Powers,* by Conrad Joyner

11. *Welfare Economics and Subsidy Programs,* by Milton Z. Kafoglis

12. *Tribune of the Slavophiles: Konstantin Aksakov,* by Edward Chmielewski

13. *City Managers in Politics: An Analysis of Manager Tenure and Termination,* by Gladys M. Kammerer, Charles D. Farris, John M. DeGrove, and Alfred B. Clubok

14. *Recent Southern Economic Development as Revealed by the Changing Structure of Employment,* by Edgar S. Dunn, Jr.

15. *Sea Power and Chilean Independence,* by Donald E. Worcester

16. *The Sherman Antitrust Act and Foreign Trade,* by Andre Simmons

17. *The Origins of Hamilton's Fiscal Policies,* by Donald F. Swanson

18. *Criminal Asylum in Anglo-Saxon Law,* by Charles H. Riggs, Jr.

19. *Colonia Barón Hirsch, A Jewish Agricultural Colony in Argentina,* by Morton D. Winsberg

20. *Time Deposits in Present-Day Commercial Banking,* by Lawrence L. Crum

21. *The Eastern Greenland Case in Historical Perspective,* by Oscar Svarlien

22. *Jacksonian Democracy and the Historians,* by Alfred A. Cave

23. *The Rise of the American Chemistry Profession, 1850–1900,* by Edward H. Beardsley

24. *Aymara Communities and the Bolivian Agrarian Reform,* by William E. Carter

25. *Conservatives in the Progressive Era: The Taft Republicans of 1912,* by Norman M. Wilensky